T0160384

Praise for *The Tragedy of American Science*

"We should read [Conner's] book as a political economy of science because science is embedded in a perverse set of cultural constraints and incentives, allowing it to be misused and manipulated in a way that endangers our democracy. Conner views science writ large, encompassing theory (disciplinary science) as well as technology. . . . The most rewarding part of the book . . . is Conner's analysis of military science since World War II. Among the scientific and technological military projects discussed by Conner, which are rarely investigated in today's popular press, are cluster bombs, Multiple Independently Targeted Reentry Vehicles, drones, cyberwarfare, the SDI, and nanotechnologies, those 'tiny insect-mimicking drones that operate in swarms, sneak into private dwelling spaces of targeted victims, and blow their heads off with microexplosive bombs.'" —**Sheldon Krimsky**, *Science, Technology & Human Values*

"American political and intellectual culture today, including scientific culture, is in a state of decay. The denial of human-caused climate change, the destruction of scientific records by the government, the attack on public education, and most recently, the Center for Disease Control's banishing words such as 'scientific-based' and 'evidence-based' are significant indications of this. The policies of the masters of corporate greed and the military-industrial complex are ruinous. We can fight back by discrediting their junk ideas and magical thinking. Cliff Conner's book helps immensely in this effort." —**Michael Steven Smith**, cohost of *Law And Disorder Radio*

"Clifford Conner's remarkable study does so much more than simply ask and answer how American science has become weaponized over the past century. *The Tragedy of American Science* is a thorough and vividly engaging account—a history of science that draws deeply on social and geopolitical analysis, and with excellently crafted case studies. It is a call to rethink the myths of American exceptionalism that, under the guise of scientific altruism and US foreign policy, have cultivated a science-for-profit system. Despite its unflinching disdain for the corporatization of research, policy, and practice, Conner's story is not a pessimistic one. Instead, with keen insight, wit, and an empathetic eye on the future, Conner helps rescue the promise of science from the tragedy it has become." —**Jacob Blanc**, author of *Before the Flood: The Itaipu Dam and the Visibility of Rural Brazil*

"Cliff Conner has brought together journalists, advocates, leakers, and litigators to restore the principles of free inquiry from its perversions by the big lies of Big Food, Big Oil, Big Pharma, and Big War. The method is true and it is simple: they lift the big rock and let fresh air and sunlight expose the little nasty, squirmy things underneath." —**Peter Linebaugh**, author of *Red Round Globe Hot Burning: A Tale at the Crossroads of Commons and Closure, of Love and Terror, of Race and Class, and of Kate and Ned Despard*

The Tragedy of
American Science

FROM THE COLD WAR TO THE FOREVER WARS

Clifford D. Conner

Haymarket Books
Chicago, Illinois

© 2022 Clifford D. Conner. First published in 2020 by Haymarket Books.

This edition published in 2022 by
Haymarket Books
P.O. Box 180165
Chicago, IL 60618
773-583-7884
www.haymarketbooks.org
info@haymarketbooks.org

ISBN: 978-1-64259-708-0

Distributed to the trade in the US through Consortium Book Sales and Distribution (www.cbsd.com) and internationally through Ingram Publisher Services International (www.ingramcontent.com).

This book was published with the generous support of Lannan Foundation and Wallace Action Fund.

Special discounts are available for bulk purchases by organizations and institutions. Please call 773-583-7884 or email info@haymarketbooks .org for more information.

Cover design by Jamie Kerry. Cover image of air force pilots bombing the southern panhandle of North Vietnam in 1966 by Lt. Col. Cecil J. Poss.

Printed in Canada by union labor.

Library of Congress Cataloging-in-Publication data is available.

10 9 8 7 6 5 4 3 2 1

To the
investigative journalists
public-interest advocates
public-interest litigators
principled scientists
whistleblowers
and
WikiLeakers
who made this book possible.

As for me and as for you,
I write for the many, not for the few.
Whether many of the many read it or not,
If wealth beyond reason is not your life's lot,
This book was most certainly written for you.

The language of Enlightenment has been hijacked in the name of corporate greed, the police state, a politically compromised science, and a permanent war economy. The economic individualism of the early, enlightened middle classes has now spawned into vast corporations which trample over group and individual rights, shaping our destinies without the slightest popular accountability. The liberal state, founded among other things to protect individual freedom, has burgeoned in our time into the surveillance state. Scientific rationality and freedom of inquiry have been harnessed to the ends of commercial profit and weapons of war.

—Terry Eagleton, *Reason, Faith, and Revolution:*
Reflections on the God Debate

CONTENTS

PREFACE TO THE 2022 EDITION

THIS BOOK WAS ORIGINALLY published in August 2020 with the subtitle "From Truman to Trump" to delineate its timeframe. Trump was voted out of office three months later, so this new edition bears an updated subtitle: "From the Cold War to the Forever Wars."

Meanwhile, a great deal of history has transpired between the summer of 2020 and the summer of 2022. A few observations are necessary to bring this "history of now" up to the present. Unfortunately, though unsurprisingly, the principal trends outlined in the first edition persist. The corporatization and militarization of American science have accelerated.

"The Bigs" Have Become Bigger Than Ever

Among all of the industries characterized by corporate gigantism, science-based Big Pharma and Big Tech have most successfully reaped megaprofits during the era of the COVID-19 pandemic. As the economic fallout from stay-at-home directives continued to wreak havoc on the livelihoods of millions of people, the *Washington Post* reported:

> The pandemic has been a boom time for America's richest billionaires. The wealth of nine of the country's top titans has increased by more than $360 billion in the past year. And they are all tech barons, underscoring the power of the industry in the U.S. economy. Tesla's Elon Musk more than quadrupled his fortune and jockeyed with Amazon's Jeff Bezos for the title of world's wealthiest person. Facebook's Mark Zuckerberg topped $100 billion. Google co-founders Larry Page and Sergey Brin gained a combined $65 billion. . . . The staggering rise in their gains contrasts with the economic devastation of millions of Americans, amid soaring unemployment and

evictions, drawing attention to issues of inequality and distribution of wealth.[1]

That was in March 2021. In January 2022, CBS News reported: "The world's nearly three thousand billionaires increased their wealth by five trillion dollars last year, a rate unprecedented in human history."[2] Meanwhile, Oxfam revealed, "Over 160 million people are projected to have been pushed into poverty."[3] Vast economic inequality continues its runaway growth, both in the United States and globally.

> *Vast economic inequality continues its runaway growth, both in the United States and globally.*

By Their Fruits Shall Ye Know Big Pharma

When the first edition of this book appeared, the COVID-19 pandemic was in its first frightening months. I had hastily added an epilogue to make note of its initial devastation. By the time the first copies had come off the presses in August 2020, the disease had claimed fewer than 170,000 lives in the United States. By April 2022 the death toll had surpassed 1,000,000.[4]

Before the onset of the pandemic, the fruits by which the corporate pharmaceutical juggernaut was known were bitter indeed. Its status in the public eye was at historical low ebb due to the half-million unnecessary deaths its opioid painkillers had caused in the United States alone. But two years later Big Pharma's reputation had risen from the ashes and shone brightly, thanks to its creation of not one, not two, but several safe and effective vaccines that promised to bring the nightmare contagion of COVID-19 under control.

The science underlying Big Pharma's COVID-19 vaccines, including the use of mRNA technology in vaccines, is impressive and commendable. The multinational industry demonstrated its capacity to accomplish remarkable scientific feats when it wants to. But there's the rub: it has to want to. And the only thing that can make it want to is the prospect of material rewards in the form of supersized profits.

That was the point of the Trump administration's Operation Warp Speed. Billions of dollars were paid out to elite Big Pharma firms like Pfizer and Johnson & Johnson as well as to venture capital startups like Moderna, Vaxart, and Novavax that had never previously brought any vaccine to market. The program was hyped as a scientific "race for a vaccine" among competing research laboratories, but in fact it was a speculative frenzy among competing hedge funds.

It took a massive infusion of government money, not private capital, to incentivize the research and development of several viable vaccines. But private investors have been allowed to reap the financial rewards, the extent of which has been deliberately hidden from public view by allowing the Warp Speed profiteers to bypass traditional federal contracting rules and regulatory oversight.[5]

Even a partial accounting, however, provides insight into the inequitable nature of the process. Moderna's vaccine, like the others, was developed with the scientific and financial support of the US government. Scientists at the National Institutes of Health contributed substantially to Moderna's success. Federal dollars financed the company's clinical trials and other research to the amount of at least $1.3 billion. To top it off, the government guaranteed Moderna a $1.5 billion market for what was at the time still an unproven product.[6] Those billions were simply the initial ante. Moderna and Pfizer together locked in guaranteed COVID vaccine sales of more than $60 billion for 2021 and 2022. Furthermore, analysts forecast 2023 revenue of more than $7.6 billion for Moderna, and they "eventually see the annual market settling at around $5 billion or higher."[7]

There was no material reason for funneling all that money through private corporations. If the pandemic was a "wartime emergency" as we were repeatedly told, why didn't the United States do what it did in World War II with the Manhattan Project and produce the vaccines in national laboratories? The Manhattan Project demonstrated that *private investment is not essential to innovation or production.* The Trump and Biden administrations, by contrast, allowed private investors to organize and control the vaccine research and production, and to become fabulously wealthy by taking all of the profits for themselves. A more rational and equitable approach would be to cut private investment out of the picture entirely by nationalizing Big Pharma.

Is that an impossible, utopian proposal? The remarkable achievements of Cuban biomedical science demonstrate that it is not. Despite Cuba's relatively modest economic status, the country has been able to mobilize its scientific and financial resources to create and produce a suite of safe, effective COVID-19 vaccines.[8] As the *British Medical Journal (BMJ)* explains, "Cuba's state-managed biotech sector has a long history of producing vaccines."[9]

Cuba was the first country to immunize its entire population against COVID-19 with its own vaccines.[10] And in sharp contrast to Big Pharma's relative neglect of poorer countries,[11]

> Cuba plans to share its vaccines with the rest of the region, much of which continues to face vaccine shortages and large outbreaks. Argentina, Mexico, and Jamaica are among those discussing potential deals, as is Vietnam. Iran has just started mass producing [Cuban vaccine] Soberana 2 after running phase III clinical trials in January.[12]

Anti-Vax Irrationalism as a Malignant Social Disease

The brief section on the anti-vax movement in chapter 5 of this book was written before the widespread political battles over COVID-19 vaccines and vaccine mandates erupted in 2021. A direct descendant of the earlier campaign against the MMR vaccine, the anti–COVID vaccine mania in the United States amplified the irrational fearmongering instigated and propagated by right-wing demagogues. Unlike the earlier agitation, however, the COVID-era anti-vax campaign included high-ranking politicians and the substantial yellow-journalism segment of the American mass media, most notably Fox News. Its loud disparagement of science and calls for resistance to vaccines as a form of civil disobedience undermined efforts to defeat the pandemic by attaining herd immunity, thus causing incalculable damage to public health, both in the United States and worldwide. The magnitude of this anti-vax campaign's criminality is evidenced by the more than six million lives the pandemic had claimed globally by the summer of 2022.

The Privatization of the Space Race:
Big Deal, Little Deal, or No Deal at All?

In July 2021, the aforementioned Jeff Bezos and another venturesome billionaire, Richard Branson, launched a privatized space race. The mass media gave it the breathless attention of a major scientific breakthrough, but what did it really amount to? Branson ascended fifty miles into the sky and Bezos soared to sixty-seven miles. According to a standard definition of "space" as beginning at the Kármán Line, about sixty-two miles above Earth's surface, Branson did not reach space while Bezos barely did. To put their accomplishments into perspective, *more than fifty years earlier* rocket science put a human being on the moon, which is more than three thousand times farther than either Bezos's or Branson's rockets traveled.

So what was the hoopla all about? "Manned" space travel has from the beginning been an exercise in misdirection designed to focus public attention on a sideshow and away from the US space program's primary purpose—the militarization and weaponization of space. And so the news media dutifully played up the dramatic competition between the plutocratic self-styled space pioneers, with all their farcical macho posturing.

Two months later, in September, a US drone attack in Afghanistan briefly put the spotlight where it belonged. The attack killed civilians, including children, but that is too common a circumstance to make it newsworthy. The reason this incident made headlines was that it came during the US withdrawal of troops from Afghanistan, which President Biden touted as "the end of the Forever War."

Does that mean this book's shiny new subtitle is already obsolete? I wish it were so, but of course it isn't. Biden's conflation of a singular Forever War with plural Forever Wars was deliberate linguistic subterfuge. More to the point, he made clear that no Forever Wars were ending by promising that US military "over-the-horizon" capabilities would "allow us to keep our eyes firmly fixed on any direct threats to the United States in the region, and act quickly and decisively."[13] An ACLU spokesperson deciphered Biden's meaning: "Airstrikes, including through the use of drones, that take place outside of recognized armed conflict—that is a centerpiece, a hallmark of the forever wars."[14]

What does this have to do with the US space program? The over-the-horizon capability to "keep eyes firmly fixed" throughout "the region" and to launch retaliatory strikes "quickly and decisively" is what the vast system of earth-orbiting satellites was created for—to provide military command and control over Earth's entire surface. Spokespersons for the weapons industry quickly piped up to remind policymakers that "more over-the-horizon operations will require resources currently not allocated. US SOCOM [Special Operations Command] and the intelligence community do not have sufficient funding to meet these new challenges."[15] Clearly, no "peace dividend" can be expected from withdrawing US troops from Afghanistan.

The Shifting Rationale for Gargantuan Military Spending

Meanwhile, with the decline of the "War on Terror" as a credible justification for endless war spending, policymakers seek to reinforce it with a rise in belligerence toward China. Although Obama's proposed "pivot to Asia" was premature, it was revived by Trump and is now being extended by Biden, whose bland denials of promoting a new Cold War are unconvincing. Although a Cold War II with China would differ in significant ways from Cold War I, it would portend a no less perilous game of chicken between countries with weapons capable of exterminating the human race.

Those who had hoped Trump's replacement by Biden would rein in the out-of-control growth of military spending were sorely disappointed. In September 2021, the House of Representatives, controlled by a Democratic Party majority, voted to approve a $777.7 billion Pentagon budget.[16] Not only was that $37 billion more than Trump's final budget, it also gave the Pentagon $25 billion more than Biden had requested! This obsequious capitulation to the military-industrial complex was so skillfully downplayed by the news media that it slipped through with hardly a ripple of public response.

But that's not all. The February 2022 Russian invasion of Ukraine prompted the generals and their lobbyists to demand *even more* war spending, and warhawk pundits and politicians launched an enthusiastic campaign to support them. The editor of the right-wing *National Review* called for a trillion-dollar military budget, while Matthew Kroenig of the Atlantic Council advised Congress that it "could go so far as to double its defense spending" without unduly burdening US resources.[17]

Similar appeals were issued by former State Department and Pentagon officials.[18] The upshot was that on March 28 the Biden administration added tens of billions of dollars to its previous military budget proposal, raising it to $813 billion.[19] The stranglehold of weaponized Keynesianism on the American economy as described in chapter 11 of this book is more evident and potent than ever.

> *The stranglehold of weaponized Keynesianism on the American economy is more evident and potent than ever.*

The danger posed by the Russian war in Ukraine and the US role in provoking it goes far beyond increased war spending. It has amplified the threat of a looming "Cold War II" beyond the US–China rivalry to include yet another armed-to-the-teeth nuclear power. Russian missile attacks on Ukrainian nuclear power plants were stark reminders of the ever-present peril of global thermonuclear Armageddon.[20]

The proximate cause of the crisis was the Russian attack on Ukraine, which human rights defenders throughout the world are duty-bound to oppose with all the courage they can muster. "The invasion of Ukraine," antiwar journalist Chris Hedges declared, "is a criminal war of aggression."[21] Nevertheless, primary responsibility for the war must be assigned to the relentless US drive to encircle Russia with threatening nuclear missile bases, culminating in the attempt to draw Ukraine into the explicitly anti-Russian NATO alliance. Calls to oppose Russian militarism with US militarism, as in demands that the US impose a "no-fly zone" over Ukraine, imperil all of humanity.

"Better with Biden"?

The November 2020 presidential election was, among other things, a very imprecise referendum on rationality. The Trump campaign luxuriated in absurd, irrational appeals to the electorate, while Biden was promoted as the champion of reason, logic, evidence, fact, and science. Viewed from that angle, it was gratifying that science won out over antiscience, albeit too narrowly to allow partisans of rationality to feel complacent. After all, more than 74 million Americans cast their votes for a pathologically narcissistic demagogue whose disdain for science was all too explicit.

Nevertheless, the positive consequences of the election outcome were far from trivial, notably with regard to public health policy and measures to combat the COVID-19 pandemic. But on the most fundamental issues, Biden's election represented a major step closer to global tragedy on multiple fronts. As noted above, in his first year in office Biden has accelerated the militarization of American science and society. And his unshakeable fealty to the fossil fuel industry draws us ever nearer to the climate catastrophe point of no return: "The Biden administration has approved 3,091 new drilling permits on public lands at a rate of 332 per month, a faster pace than the Trump administration's 300 permits per month." Furthermore, Biden "recently opened more than 80 million acres in the Gulf of Mexico to auction for oil and gas drilling, a record offshore sale that will lock in years of greenhouse gas emissions."[22]

So is the world "Better with Biden" than if Trump had won reelection? That is an extremely low bar. Although a strong case can be made that a second Trump term could have resulted in a rapid "Pinochetization" of the United States, the answer nonetheless is "no." With Biden no less than Trump, the planet Earth is precipitously hurtling into the abyss.

With Biden no less than Trump, the planet Earth is precipitously hurtling into the abyss.

Good News/Bad News on the Climate Crisis

The bad news is that over the past two years, whatever steps might have been taken to decrease atmospheric CO_2 weren't. Two more precious years were wasted, and the existential threat to humanity has intensified. The good news is that a tectonic shift is underway in the public discourse on the subject, away from denial and toward awareness. That at least opens the door to the possibility of a Great Awakening that could conceivably result in effective action to forestall the approaching disaster.

The price of that nascent enlightenment has been steep. The two years of wildfires, hurricanes, floods, droughts, and other extreme weather events that have forced widespread acknowledgement of the danger have cut short many lives and ruined many more.

Blockchain Bamboozlement and the Bitcoin Bubble

The impending climate catastrophe requires attention to another urgent sci-tech issue. Cryptocurrencies like Bitcoin and their underlying blockchain technology are tech innovations that the original edition of this book did not discuss. Recent experience has confirmed early worries that they are a very, very bad idea indeed.

Most disturbing is the outrageously large carbon footprint that accompanies their production. Bitcoin and other cryptocurrencies are "mined" by entire warehouses full of constantly running specialized computers. The cooling power to keep them from overheating consumes an inordinate amount of energy, mostly in the form of coal-generated electricity. A February 2021 BBC report revealed that Bitcoin production uses more than 121 terawatt-hours of electricity per year—more than the entire country of Argentina consumes.[23] And that number is growing continually with no end in sight.

What value does this unconscionable expenditure of energy produce? Bitcoin has no material value. A commonplace description of the material value of a dollar bill puts it at "no more than the paper it's printed on." Bitcoin is not printed on paper, so its material value is even less.

Meanwhile, the *price* of a single Bitcoin has explosively inflated, rising from essentially zero at its inception in 2009 to tens of thousands of dollars today. That price is determined by buyers bidding for them in a manipulated market. It is analogous to the international art market, in which a painting merely *thought to be* by Leonardo da Vinci can sell for $450 million.[24]

And all for what? Bitcoin's only "use value" is as a currency that is crypto—which is to say, hidden from public accountability. Most news reports have focused on Bitcoin holders' most unsavory motives for hiding money, such as to finance illegal drug dealing or to collect ransomware payments. But the essential motivation driving cryptocurrency is old-fashioned tax evasion. Libertarians' fever dreams of freeing currency from the clutches of tyrannical government are simply cover for the desire to hide their wealth from tax authorities. As facilitators of ever-growing global economic inequality, cryptocurrencies are in their very essence inimical to the public interest.

There is an even darker underside to the crypto story. Aside from its use as a money laundering and asset hiding medium, it is luring gullible small investors into a gigantic decentralized Ponzi scheme. As one astute critic explains, cryptocurrency is "the people's Ponzi." Unlike Bernie Madoff's fraudulent scheme, which swindled wealthy investors, when crypto crashes, "it is everyday working people who will suffer most when their savings inevitably evaporate."[25]

The blockchain algorithm on which Bitcoin runs is an ingenious technological innovation. It has proven to be useful, for example, in rural China where it gives consumers a trustworthy means of assuring food safety that the Chinese government is unable to provide.[26] A socialist commentator has argued that "blockchain technology can serve as a mechanism for worker cooperatives and the Left."[27] Sadly, however, whatever potential blockchain has to improve the human condition is currently being grossly misused in a way that threatens humanity's very existence.

The Tragedy—and Thus the Struggle—Continues

The conclusion of this book has not changed in the two years that have passed since its original publication. The tragedy of American science, and by extension of human society on the planetary scale, has only worsened. I repeat: "The need to replace the global economic system that serves private interests with one that serves the public interest is more urgent than ever." Down with science for corporate profit maximization! Science for the people!

Clifford D. Conner
New York City, on the land of the Lenape Nation
April 2022

FOREWORD

THIS IS A WORK OF contemporary history. The mercurial nature of current American politics makes writing this form of history more daunting than usual. From the time the manuscript is submitted to the publisher to when the book comes off the presses, some details may already be out of date. Unfortunately, however, the essential theme of the book is unlikely to be outdated anytime soon. How I wish that were not so!

The sources of information upon which I have drawn are likewise contemporary rather than archival. For the most part, they originated not with scientists but with investigative journalists, public-interest advocates and litigators, conscientious leakers, and whistleblowers (some of whom have also been courageous scientists). That is because the tragedy of contemporary science is less about science than about economics, politics, and public relations. The scientists did not cause it; many of them, too, are its victims.

The title is a conscious echo of one that strongly influenced my own path of development as a historian: William Appleman Williams's *The Tragedy of American Diplomacy*.

The book originated as a chapter I contributed to a collection of essays entitled *Imagine: Living in a Socialist USA* (HarperCollins, 2014). In imagining how the practices of science and technology could be optimized, I was forced to confront their current state. I found them not only deficient but also tragic, which led me to transform that essay into the book you are now reading.

A Coronavirus Postscript
As if to illustrate my caveat that rapidly changing circumstances might overtake what I am writing here, the outbreak of the COVID-19 pandemic issued an unprecedented challenge to American science just as

the book was about to go to press. Although this story has only begun to unfold, its enormous impact and potential consequences demand an initial assessment of what it means for science and society alike. I have attempted to provide one in an epilogue to this book.

Clifford D. Conner
May 2020

INTRODUCTION

I AM ILL SUITED FOR the role of Jeremiah. I have a naturally sunny disposition. Mr. Bluebird's on my shoulder.

There is a scientific explanation for that: I am a bearer of the "feel-
good gene." A professor of psychiatry at Weill Cornell Medical College explains that I (and about 20 percent of the population of the United States)
"won the genetic sweepstakes" due to a fortunate genetic mutation that
"produces higher levels of anandamide—the so-called bliss molecule and
our own natural marijuana—in our brains."[1] This propensity for happiness can be overridden by sorrowful real-world intrusions such as deaths
of friends and loved ones, but the predisposition toward joy persists.

On top of that, I am a science-and-technology junkie and have been at
least since I passed the test for a ham radio license—call letters K4EJO—
at age fourteen. My enduring fascination with the subject led me to become a historian of science. So it actually pains me to be the bearer of bad
tidings who publishes a jeremiad entitled *The Tragedy of American Science*.
Real-world problems have overwhelmed my feel-good gene. But not entirely; ever the optimist, I have ended this book on a hopeful note.

Describing the state of science as tragic would have seemed peculiar
to most people as recently as the first half of the twentieth century. Science's reputation was then golden. The expectation that modern science
could, and would, solve all of humanity's problems was almost universal
among educated people.

That happy pipe dream received a double jolt during World War II.
First came the horrors of Nazi racial science and its accompanying technology of human extermination. That was followed by the instant incineration of more than a hundred thousand inhabitants of two Japanese
cities, marking the advent of the nuclear age. J. Robert Oppenheimer,

1

one of the atomic bomb's creators, quoted a line from a Hindu holy book to signal the emergence of science's ominous dark side: *Now I am become Death, the destroyer of worlds.*

The Manhattan Project's success in producing nuclear weapons transformed American science into Big Science. The ensuing Cold War fed policymakers' paranoiac obsession with the technology of death. The out-of-control proliferation and use of weapons of mass destruction is the worst of contemporary science's tragic fruits, but there are many others. The misuse and abuse of science to justify destroying the earth's habitability has also become a source of widespread anxiety.

But Is It Really a Tragedy?

The idea that anything about American science could be tragic may rankle some readers. Isn't the United States where most of the world's technological innovations have been and still are produced? Airplanes? Television? The computer? The internet? The iPhone? Hasn't American science also been responsible for great medical and biochemical advances? Its current state may present some difficulties and challenges, but isn't it hyperbole to call it a *tragedy*?

Unfortunately, no. The river of the tragedy has two headwaters: corporatization and militarization. Both are consequences of a profit-driven economic system that hamstrings humanity's ability to make rational economic decisions.

The Corporatization, or Privatization, of Science

Science is presumed to be trustworthy because it is based in objective fact rather than subjective bias. By definition, that requires research to be conducted impartially, by scientists with no conflicts of interest that could affect their judgment. But science propelled by private profit cannot avoid material conflicts of interest that are anathema to objectivity.

As corporate domination of science and technology has grown, the ideal of objective scientific investigation has fallen by the wayside. So-called scientific studies are now routinely conducted by individuals and institutions with large financial stakes in the outcomes. ExxonMobil and Koch Industries fund climate-change-denial studies.[2] Philip Morris and R. J. Reynolds produce findings minimizing the link between smoking and lung cancer.[3] Purdue Pharma and Pfizer investigate the benefits and risks of the medicines they sell.[4] Coca-Cola and Kellogg enlist nutritional science as a marketing tool for their products.[5]

A rapidly growing engine of privatized science is Big Tech, also known as the Big Five—the technology corporations that dominate the computer, software, and online services industries. Google, Apple, Facebook, Amazon, and Microsoft are pumping billions of dollars into creating "machine intelligence" that can do everything from driving trucks and cars to performing brain surgery. Tech analyst Farhad Manjoo writes that the Big Five are collectively

> among the biggest investors in research and development on the planet. According to their earnings reports, they are on track to spend more than $60 billion this year [2017] on research and development. By comparison, in 2015, the United States federal government spent about $67 billion on all nondefense-related scientific research.[6]

The intended result is an artificial-intelligence future that is being designed to serve private rather than public, and military rather than civilian, interests.

The scientific method has been refashioned to fit the new reality. "Hypothesis-driven research" now signifies formulating propositions to advance corporate agendas and designing studies to provide evidence aligned with them. Investigations that produce a semblance of support for a desired hypothesis are accorded full public relations treatment, while those that do not are quietly consigned to the paper shredder.

The results of all this research are at worst fraudulent and at best untrustworthy. The most egregious examples are the voluminous climate-change-denial and smoking-doesn't-cause-cancer studies. Yet, abetted by legislators and a credulous mass media, corporate science-for-profit shapes the public discourse and public policy that adversely affect our environment and our health.

The Institutionalization of Science-for-Profit

Rational voices such as those of Naomi Oreskes, Erik Conway, Sheldon Krimsky, and Marion Nestle[7] have raised concerns and warnings about the corruption of science by Big Money, but the voices serving corporate interests in the major media and right-wing think tank universe[8] are louder. By skillful application of the false-equivalence fallacy, the latter have managed to skew the conversation far to the irrational side.

The manipulation of research results to serve private commercial interests is anti-science masquerading as science. The forces of antiscience have strong institutional support in the United States, most significantly in our fundamental political institutions. From federal and state legislatures to local school boards, demagogic politicians have spearheaded what has been called by some "the Republican war on science." While the loudest of the know-nothings have indeed been Republicans, anti-science has found enablers on both sides of congressional aisles, funded by bipartisan corporate money.

Enter Trump

By this point, many readers must be thinking: "When does Trump come into this discussion?" Is Donald Trump's administration not the purest expression of the tragedy of American science?

Of course it is. Donald Trump is by no means a run-of-the-mill Republican politician. Before Trump, it would have been difficult to imagine the cautious *New York Times* saying of an American president, "Mr. Trump is not just racist, ignorant, incompetent and undignified. He's also a liar."[9] Innumerable pundits have identified Trump's inauguration as the start of a "post-truth" or "post-fact" era, an anti-Enlightenment, an Age of Unreason. It has created an unsettling mindscape in which consensus is fragmented, the bedrock of discourse is crumbling, and the pursuit of reliable knowledge is fading into irrelevance. The domination of the national conversation by clowns and charlatans has put science under siege in the United States.

However—and this is crucial—the tragedy of American science did not begin with the election of Donald Trump. The blustering demagogue's assault on science, empirical reality, and rationality brought pre-existing trends to the surface and made them more prominent, but they had been incubating since the end of the Second World War. Trump's ideological lineage prepared him perfectly as a continuator: his mentor, Roy Cohn, had in turn been mentored by the archetypal Cold Warriors Joseph McCarthy and J. Edgar Hoover.

A particularly dangerous science trend that the Trump administration has exacerbated is the weakening of governmental regulation of commercial activities that contaminate the air we breathe, the water we drink, the food we eat, and the medicines we take. The ability of the Environmental Protection Agency and the Food and Drug Administration to provide even the limited oversight they are expected to provide has significantly declined.

Large sections of the American intelligentsia—"the best brains money can buy"—have also been enlisted in the effort to undermine scientific integrity. Many honest scientists struggle mightily to retain independent control over their research, but as public funding has retreated, the pressure to succumb to the quest for private funding has continuously increased. As chapters 8 and 9 will demonstrate, university laboratories and think tanks frenetically compete for corporate sponsors, and the latter have been happy to oblige. The upshot is that science in the public interest is largely a thing of the past in the United States. It is now science in the corporate interest.

The Militarization of American Science

The most tragic distortion of American science, in my opinion, is the consequence of its extreme militarization. Since the end of World War II, the primary mission of American science has been to create new and more efficient ways to kill people—from thermonuclear bombs to robot drones, cluster munitions, and antipersonnel weaponry of seemingly infinite variety.

Big Science literally exploded onto the scene as a result of the Manhattan Project during World War II. Its success was validated by the mushroom clouds that rose over Hiroshima and Nagasaki in August 1945, thus establishing the martial inclination of Big Science's development during the Cold War and beyond.

Science and technology, rather than being the creative engines of human progress, have instead been reoriented toward destructive and antihuman ends. The proof is in the US budget, where more than half of the Research and Development funding—amounting to trillions of dollars over the past decades—has gone for military purposes. In 2017, 48.5 percent of federal R&D spending went to the Department of

Defense alone, and that doesn't include the vast expenditures on nuclear research in the Department of Energy's budget.[10]

American science has also produced beneficial civilian technologies, but most of them have been secondary byproducts of war technology. In 1946, for example, a Raytheon engineer observing radar tube tests felt a strange warmth in his pocket. Realizing that emissions from the tubes had melted a candy bar he was carrying, he investigated the phenomenon further, and the eventual result was the microwave oven.[11] Other prominent offshoots of military research ("dual use" technologies) include the internet, GPS, and "virtual assistants" such as Alexa and Siri.

Imagine, by contrast, what could be accomplished if all of that money and all of that scientific talent were instead directed toward finding solutions to the crucial problems facing the human race today—poverty, hunger, disease, and environmental devastation. But they are not, and if that isn't a tragedy, the word holds no meaning.

The military chokehold on the federal budget and its science component has been downplayed in the public discourse. President Eisenhower's 1961 warning about the dangers of the "military-industrial complex" is well known, but the conversation went no further. Coming to grips with it requires understanding the Cold War history of American science, from the Truman administration to the present, and how it was shaped by Operation Paperclip, the RAND Corporation, and DARPA, the Defense Advanced Research Projects Agency.

Most distressing is the realization that this problem cannot be fixed—at least not in the present US context. That is due to the US economy's dependence on extremely high levels of military spending. The fatal malignancy has a name, "weaponized Keynesianism," which sounds terribly academic but is useful as shorthand to identify the problem. Here is its essence: without the trillion-plus dollars a year in military spending, the unemployment rate would rise to the skies.

In the United States today, if the Pentagon ceased to provide a gigantic artificial market for industrial production, millions of workers—and not only weapons-plant workers—would lose their livelihoods. Without paychecks they would be unable to buy goods or services and the wheels of the national economy would grind to a halt, which would also massively disrupt the global economy.

American Exceptionalism

American Exceptionalism includes the contention that the United States is not bound by traditional norms of morality in international affairs. Invading other countries and killing their inhabitants when they resist would generally be considered war crimes. The use of torture has been condemned as immoral and repugnant by all civilized peoples since the eighteenth-century Enlightenment. But when the United States invades, kills, and tortures, it is deemed acceptable by administrations and citizens alike because the United States is assumed to be a benevolent superpower that acts only in defense of peace, democracy, and human rights.

This normalizes the outrageously inflated US arms production as being for a good cause. It serves as the rationale for a national security state that monitors the private communications of everyone in the country and spawns secret terrorism tribunals that dangerously erode the rule of law. But because American Exceptionalism has served as the ideological justification for many wars, resulting in the death of many millions of people all over the globe, a critical rethinking is long overdue.

But All Is Not Lost! (Breaking Out of the Box)

Does the preceding litany of woe imply unavoidable doom? That is not my intent. The current tragedy of American science may appear to lead to a hopeless impasse, but there is a way out.

It is not an easy way out. To comprehend it requires some serious thinking outside the box. In this case, "the box" is the market-based, profit-driven economic system that almost all American commentators and ideologues take for granted, as if no alternative system is worth mentioning or even possible.

This has, for many decades, been a strong, indestructible box that has successfully imprisoned the minds and constrained the thinking of almost the entire American public. But the 2016 elections began, perhaps, to reveal stress fractures in the box. The campaign of Bernie Sanders brought the word *socialism* into the public discourse for the first time in most Americans' living memory.

I do not believe Bernie Sanders's 2016 candidacy offered a solution to the crucial conundrum of weaponized Keynesianism. Despite his personal opposition to military spending and war, the American congressional political system is structured to ensure the futility of any and all challenges to the war machine. The record of Sanders's quarter century as senator and congressman from Vermont reveals that, while criticizing "excessive" Pentagon budgets, he frequently voted to approve them,[12] and to approve military aggression in Iraq, Afghanistan, Kosovo, Somalia, and elsewhere, demonstrating that he posed no real threat to the warhawks' agenda.[13]

However, the 2016 Sanders campaign and its 2020 sequel deserve a great deal of credit for popularizing and destigmatizing the idea of socialism among young Americans. At least the way is now open for serious discussion of alternatives to the way the American economy is currently organized. The 2018 elections furthered the process as self-proclaimed democratic socialists began to win office on the national, state, county, and municipal levels. The run-up to the 2020 elections also stimulated public discussion of science-related issues under the rubrics "Green New Deal" and "Medicare for All." The Green New Deal in particular reflected a growing consciousness among young people that global warming is an existential crisis. Avoiding that catastrophe, moreover, implies the need for massive structural social and economic change rather than mere tweaking-the-system reforms.

One element of the general discussion is whether science and technology can be reoriented from destructive purposes to creative ones. Fortunately, history offers some compelling indications that genuine, unfettered science could thrive in a postcapitalist economic order. The concluding chapter will examine this history and demonstrate that scientific advance is not absolutely dependent upon individual material incentives and the profit motive.

The hopeful note in all of this is that replacing the current science-for-profit system with a science-for-human-needs system is not an impossible, utopian dream.

The
Corporatization
of
American
Science

CHAPTER 1

The Big Fat Lie

RESEARCH INTO FOOD-RELATED problems is conducted by scientists trained in a large number of traditional disciplines, from microbiology and physical chemistry to gastroenterology and psychiatry. This inquiry into the current state of the food sciences, however, focuses on two broader areas of interdisciplinary interest: nutrition science (the subject of this chapter) and agronomy (the subject of the next chapter). Nutrition science concerns the interaction of food and human bodies, and agronomy has to do with how food is produced.

Is Nutrition Science an Oxymoron?

The crucial role of nutrition in human health means that everyone has a stake in the accuracy of information about the food we eat. Almost half the adult population of the United States—133 million people—has at least one of the four leading diet-related health problems: heart disease, stroke, cancer, or diabetes. Those chronic conditions are responsible for seven out of ten American deaths every year.[1]

My personal experience with food-related advice—a lifelong effort to keep my weight under control—has led me to think of the phrase "nutrition science" as an oxymoron. However unfair that characterization may seem, frequent scandals have laid bare significant ways in which nutrition research has indeed been deeply corrupted.[2]

In September 2018, the career of "one of the most respected food researchers in America . . . came to an unceremonious end," wrote Anahad O'Connor in the *New York Times*. Dr. Brian Wansink, head of the Food and Brand Lab at Cornell University, resigned after a university investigation found him guilty of "academic misconduct in his research and scholarship, including misreporting of research data."[3]

Some experts perceived this scandal to be "symptomatic of a broader problem in food and health research." Critics "argue that an alarming number of food studies are misleading, unscientific or manipulated to draw dubious conclusions."

> Dr. Wansink's lab was known for data dredging, or p-hacking, the process of running exhaustive analyses on data sets to tease out subtle signals that might otherwise be unremarkable. Critics say it is tantamount to casting a wide net and then creating a hypothesis to support whatever cherry-picked findings seem interesting— the opposite of the scientific method.[4]

Dr. Wansink's disgrace and the exposure of his unscrupulous methods are, unfortunately, not expected to lead to major improvements in nutrition science practices. As will become clear in the examples below, studies based on flimsy statistical evidence are far from rare, and no one thinks data dredging is going away anytime soon.

How Scientific Are the Government's Dietary Guidelines?

If you had come to accept as an article of faith that eating low-fat foods is essential to the good health of your heart, you were not alone. Meanwhile, public awareness of sugar as a factor in causing heart disease has been comparatively nonexistent. These perceptions had long been promoted as scientifically based by the nutrition profession and bolstered by the stamp of government approval.

The federal government issues and updates its *Dietary Guidelines for Americans* every five years. Whether or not very many Americans pay direct attention to them, the guidelines have a substantial impact on the way we eat. They form the basis of nutrition education, food labeling laws, food-assistance programs, and research priorities at the National Institutes of Health. Those food-assistance programs, by the way, directly impact a quarter of the American population. School breakfast and lunch programs and the food-stamp program SNAP (Supplemental Nutrition Assistance Program) expend more than $100 billion a year to provide food that is legally bound to meet the federal nutrition requirements. Furthermore, the guidelines affect not only Americans; they serve as a model for governmental dietary recommendations throughout the world.

The guidelines are required by law to reflect the "preponderance of the scientific and medical knowledge" regarding nutrition.[5] The issuance of the most recent set, in 2015, was preceded by a preliminary report authored by a Dietary Guidelines Advisory Committee of appointed experts.[6] Their report was then to be shaped into the official *Dietary Guidelines for Americans* and released under the aegis of the US Department of Agriculture and the Department of Health and Human Services.[7]

When the preliminary report appeared in February 2015, it attracted an unprecedented level of interest. Whereas the previous report, five years earlier, received only about two thousand public comments, this one was greeted by twenty-nine thousand. And that was but the opening round of raging debates between consumer advocates and lobbyists over sugar, red meat, sustainable agriculture, and science. The *BMJ* (formerly the *British Medical Journal*) published an in-depth analysis of the report, sharply criticizing it as insufficiently scientific.[8]

A key element of the *BMJ*'s critique was that the science allegedly underlying the report was tainted by financial conflicts of interest. The committee of experts acknowledged that they had relied heavily on data provided by health advocacy organizations funded by the food and pharmaceutical industries.

The American Heart Association, for example, had received "decades of support from vegetable oil manufacturers, whose products the AHA has long promoted for cardiovascular health."[9] Furthermore, members of the expert committee itself had undisclosed conflicts of interest. One had received research funding from walnut growers and vegetable oil giant Unilever; another from publicists for General Mills and PepsiCo; and for the first time, the committee chair was not from a university but from a health-industry corporation. In summary, the *BMJ* charged, "This reliance on industry backed groups clearly undermines the credibility of the government report."

Nutrition Science's Big Sugar Addiction

A 2016 investigative report by Anahad O'Connor in the *New York Times* cited newly discovered documents demonstrating that "five decades of research into the role of nutrition and heart disease, including many of today's dietary recommendations, may have been largely shaped by the sugar industry."[10] The evidence is summarized in a 2016 article published in the American Medical Association's *JAMA Internal Medicine.*[11] Internal documents of the sugar industry reveal that an industry trade group named the Sugar Research Foundation (later renamed the Sugar Association) funded a 1967 study that set the agenda for nutritional research and provided the propaganda message for Big Sugar for decades to come.

Three Harvard professors were paid by the Sugar Research Foundation to produce a review[12] of research studies that had been chosen by the trade group itself. Their paper, which appeared in July 1967 in the authoritative *New England Journal of Medicine,*[13] minimized the role of sugar in heart disease and implicated saturated fat instead. The thousands of pages of archival documents that came to light leave no room for doubt about the conscious nature of the collusion between the Harvard scientists and the sugar industry:

> [In 1964] John Hickson, a top sugar industry executive, discussed a plan with others in the industry to shift public opinion "through our research and information and legislative programs." At the time, studies had begun pointing to a relationship between high-sugar diets and the country's high rates of heart disease. . . . Mr. Hickson proposed countering the alarming findings on sugar with industry-funded research. "Then we can publish the data and refute our detractors," he wrote.
>
> In 1965, Mr. Hickson enlisted the Harvard researchers to write a review that would debunk the anti-sugar studies. He paid them a total of $6,500, the equivalent of $49,000 today. Mr. Hickson selected the papers for them to review and made it clear he wanted the result to favor sugar. Harvard's Dr. Hegsted reassured the sugar executives. "We are well aware of your particular interest," he wrote, "and will cover this as well as we can."[14]

The corruption of this science cannot be dismissed as inconsequential. Historian of science David Singerman concluded that it "helped set the agenda for decades of public health policy designed to steer Americans into low-fat foods, which increased carbohydrate consumption and exacerbated our obesity epidemic."[15]

Sugar-Water and Candy Science

The Coca-Cola Company has been particularly culpable in its efforts to misdirect nutrition science regarding sugar. In 2014 it funded the creation of a front group, the Global Energy Balance Network, to refute critics who identify Coca-Cola and other sugary beverages as prime culprits responsible for the burgeoning epidemic of childhood obesity and rise of type 2 diabetes. A 2001 Surgeon General's report had declared: "Overweight and obesity have reached nationwide epidemic proportions." The rates have continued to climb.

How burgeoning is the epidemic? In 1980 fewer than 15 percent of Americans were obese; today more than two-thirds of adults and a third of children and adolescents are obese or overweight. As for type 2 diabetes, which was rare among children in 1980, according to current estimates increasing numbers of adolescents and fully half of all adults have the disease or are prediabetic.

The Global Energy Balance Network's now defunct website previously claimed to be "the voice of science" with regard to obesity research. The science it was promoting asserted that children get fat not due to excess caloric intake but because they do not exercise enough. The front group was designing its own studies and cherry-picking data to back up that bogus claim.

Another investigative report by Anahad O'Connor revealed that Coca-Cola "has teamed up with influential scientists who are advancing this message in medical journals, at conferences and through social media."[16] When the report exposed the Global Energy Balance Network as a Coca-Cola sock puppet, the website promptly disappeared. At the same time, Coca-Cola announced the "retirement" of its chief scientist, Dr. Rhona Applebaum, who had presided over the public relations debacle.

One of the scientists with whom Dr. Applebaum had collaborated was Dr. James O. Hill, a professor and obesity researcher at the University of Colorado medical school. In an email pitching a research program to a Coca-Cola executive, Dr. Hill demonstrated more salesmanship than scientific acumen:

> Here is my concept. I think it could provide a strong rationale for why a company selling sugar water SHOULD focus on promoting physical activity. This would be a very large and expensive study but could be a game changer. We need this study to be done.[17]

In a later email, Dr. Hill added, "I want to help your company avoid the image of being a problem in peoples' lives and back to being a company that brings important and fun things to them."[18] That proposal was dated November 9, 2015. Six weeks later, the *Denver Post* reported that Hill had received $550,000 from Coca-Cola for an "obesity campaign."[19]

Dr. Steven N. Blair, an obesity researcher at the University of South Carolina, received $3.5 million from Coca-Cola between 2008 and 2015 to shift the blame from sugary soda to a lack of exercise. Dr. Blair's highly influential research "has formed much of the basis of federal guidelines on physical activity."[20] Nevertheless, the nutrition department at Harvard and thirty-six cosigning scientists described Dr. Blair's claims as "scientific nonsense."[21] Blair, however, unfazed by the criticism, was quoted in the *Atlantic* as saying, "Don't ask me to list all the corporations that I've received some consulting fee or honorarium from. It would probably take me days to go back through mountains of computer files to find them."[22]

Three years after Coca-Cola shut down the Global Energy Balance Network, the company was found to be running the same scam under a different name in China:

> Happy 10 Minutes, a Chinese government campaign that encouraged schoolchildren to exercise for 10 minutes a day, would seem a laudable step toward improving public health in a nation struggling with alarming rates of childhood obesity.[23]

"China's fitness-is-best message," however,

> has largely been the handiwork of Coca-Cola and other Western food and beverage giants, according to a pair of new studies that document how those companies have helped shape decades of Chinese science and public policy on obesity and diet-related illnesses like Type 2 diabetes and hypertension.[24]

Further investigation into this affair revealed the hidden hand of the International Life Sciences Institute, a "shadowy group" that "shapes food policy around the world." The ILSI, a US front group funded by four hundred corporate members including Coca-Cola, PepsiCo, General Mills, and DuPont, has branches in seventeen countries and has been especially influential in China, India, and Brazil.[25]

Meanwhile, back in the United States, another significant recipient of Coca-Cola research money was Dr. Brenda Fitzgerald, whom the Trump administration appointed in July 2017 to head the Centers for Disease Control and Prevention, a key federal science post. In her previous job as health commissioner of the state of Georgia, Fitzgerald had accepted a million-dollar grant from Coca-Cola to study childhood obesity. She was forced to resign as CDC director in early 2018 due to another conflict-of-interest scandal involving her investments in the tobacco industry.

Coca-Cola is far from alone in its attempt to wrap its business interests in the mantle of science. An Associated Press investigation into the food industry's influence on food policy uncovered thousands of incriminating emails via Freedom of Information Act requests for public university records. One revelation involved an allegedly scientific finding that "children who eat candy tend to weigh less than those who don't."[26] The funding for the study that produced that paradoxical gem came from the National Confectioners Association, a trade association representing the people who bring us Butterfingers, Hershey chocolates, and Skittles. The study's authors were two professors of nutrition, Carol O'Neil of LSU and Theresa Nicklas of the Baylor College of Medicine, and Victor Fulgoni, a former Kellogg executive who now, according to the Associated Press, devotes himself to helping companies advance "aggressive, science-based claims about their products." These three have been quite prolific; since 2009 they "have written more than two dozen papers funded by parties including Kellogg and industry groups for beef, milk and fruit juice."[27]

Nutrition Science versus "Nutritionism"

Creating front groups and paying scientists to tailor their research to serve food industry interests are direct methods by which nutrition science has been corrupted. More subtle strategies for deceiving the public have also been devised. One particularly successful stratagem has been to shift the focus of nutrition research from foods to molecules.

"For 150 years," historian of science David Singerman explains, the sugar industry "has shaped government policy in order to fuel our sugar addiction."[28] In the nineteenth century, the American sugar industry launched a two-pronged attack to protect its home market from foreign competition. The first prong was a lobbying effort for higher import tariffs on sugar. "As a result, the government got hooked on sugar, too: By 1880, sugar accounted for a sixth of the federal budget."[29] The second prong was a push for legislation to put US sugar's Caribbean competitors at a disadvantage by having "refined sugar" legally defined according to measurement of its sucrose content. This sensible-sounding appeal to scientific objectivity, Singerman says, "masked nefarious aims":

> Like the tobacco industry in the 1960s, these refiners knew that scientific questions were hard for outsiders to adjudicate, and thus easier to manipulate to an industry's advantage.[30]

The sleight of hand that shifted attention from sugar to sucrose continues to the present day to bend nutrition science toward the interests of Big Food in general:

> Looking back at the industry's transformation of sugar (an edible substance derived from a plant) into sucrose (a molecule), we also see the roots of "nutritionism" in United States policy. That's the idea that what matters to human health is not food per se but rather a handful of isolable biochemical factors.[31]

"Nutritionism," public-interest food critic Michael Pollan explains, "is not quite the same as nutrition. As the 'ism' suggests, it is not a scientific subject but an ideology." The guiding premise of nutritionism is that the proper focus of a science of food is the nutrient. And because "nutrients, as compared with foods, are invisible and therefore slightly mysterious," we need scientists "to explain the hidden reality of foods to us."[32]

The concept of nutrients originated early in the nineteenth century, with the identification of protein, fat, and carbohydrates as "macronutrients." Toward the end of the century a biochemist, Kazimierz

Funk, discovered substances he named "vital amines," or "vitamines" for short, one of which gave protection against the deadly disease beriberi, and thus were born the first micronutrients. However, it was not until the late 1970s that nutritionism emerged as a dominant paradigm. That resulted from a food industry backlash against dietary guidelines proposed in 1977 by a US congressional committee—popularly known as the McGovern Committee—advising Americans to consume less red meat and dairy products.[33] The committee had originated in 1968, and its pioneering *Dietary Goals* report represented several years of research and debate on the part of government and university scientists.

It is axiomatic that no industry wants consumers to consume less of its products. By 1982 a ferocious lobbying campaign by the cattle, dairy, egg, and sugar industries had succeeded in having the "eat less" recommendation replaced by a less threatening alternative: "Choose meats, poultry and fish that will reduce saturated-fat intake."[34]

Notice, Pollan points out,

> how distinctions between entities as different as fish and beef and chicken have collapsed; those three venerable foods, each representing an entirely different taxonomic class, are now lumped together as delivery systems for a single nutrient. Notice too how the new language exonerates the foods themselves; now the culprit is an obscure, invisible, tasteless—and politically unconnected—substance that may or may not lurk in them called "saturated fat."[35]

From that time forward, government dietary guidelines would be "clothed in scientific euphemism and speaking of nutrients." When a landmark National Academy of Sciences (NAS) report on cancer and nutrition appeared in 1982,

> it codified the official new dietary language. Industry and media followed suit, and terms like polyunsaturated, cholesterol, monounsaturated, carbohydrate, fiber, polyphenols, amino acids and carotenes soon colonized much of the cultural space previously occupied by the tangible substance formerly known as food. The Age of Nutritionism had arrived.[36]

Reducing food to nutrients is a reductionist fallacy that Pollan skillfully deconstructs. "Even the simplest food," he explains, is a hopelessly complex thing to study, a virtual wilderness of chemical compounds, many of which exist in complex and dynamic relation to one another, and all of which together are in the process of changing from one state to another.[37]

A major problem with reductionist science is that it encourages mechanistic interpretations: "Put in this nutrient; get out that physiological result." That ignores not only the physiological differences between individuals but also differences between entire cultures and societies:

> Some populations can metabolize sugars better than others; depending on your evolutionary heritage, you may or may not be able to digest the lactose in milk. The specific ecology of your intestines helps determine how efficiently you digest what you eat, so that the same input of 100 calories may yield more or less energy depending on the proportion of Firmicutes and Bacteroidetes living in your gut. There is nothing very machinelike about the human eater, and so to think of food as simply fuel is wrong.[38]

The story of nutritionism illustrates how corporate influence has discouraged research into entire categories of food products. That is but one way the integrity of food science has been compromised—and all of them threaten the physical well-being of the American public. Warning: corrupt science can be detrimental to your health.

CHAPTER 2

The Green Revolution

THE GREEN REVOLUTION HAS for more than seven decades been an on-again, off-again attempt to eradicate, or at least alleviate, the scourge of world hunger. The magnitude and urgency of the problem, though not always at the center of our consciousness, are obvious. Malnutrition underlies diseases that kill more than three million children under the age of five every year.[1]

Childhood hunger does not exist only in underdeveloped countries; it is a significant problem in the United States as well. Infant death rate statistics reflect a society's food insecurity; in 2016 the CIA's *World Factbook* ranked the United States fifty-seventh among the world's countries in infant mortality, just behind Bosnia and Herzegovina. The questions this poses are straightforward and simple to understand: Why can't everyone in the world have enough to eat? Can agronomy—the science of food production—provide a solution to the problem?

Malthus's Answer
In 1798, Reverend Thomas Malthus, in his *Essay on the Principle of Population*, advanced a profoundly erroneous reply to the first question. His answer was that the question has no answer: everyone *cannot* have enough to eat, because poverty and starvation are "nature's way" of controlling population. A veneer of mathematical hocus-pocus—the claim that population increases geometrically while food production increases only arithmetically—lent the apparent authority of science to his pronouncements, thereby increasing their appeal to social theorists and politicians.

The deeply reactionary social consequences of Malthus's assertions soon became clear. In 1834 the English government introduced pitiless legislation inspired by Malthusianism to justify ending relief payments to its poorest citizens, forcing them into brutal workhouses. The social philosopher Herbert Spencer misapplied Darwin's theory of natural selection to create Social Darwinism, providing the ultimate justification for the callousness and rapacity of unfettered capitalism. Spencer erroneously maintained that Darwinian "survival of the fittest" operates not only in biological evolution but also in human society.[2] In its harshest form, Social Darwinism suggested that the human race would be best served by allowing the poor to perish by starvation rather than helping them survive by means of private charity or state-sponsored welfare programs. Although population control by famine never became official British policy, Spencer's ideas reinforced the political tendencies that blamed and punished the poor for their poverty.

As for Reverend Malthus's 1798 prophecy that population growth would forever eclipse our planet's capacity to provide food enough for all, his timing could not have been worse. No sooner had he made that pronouncement than the Industrial Revolution began to take off, resulting in a rapid, sustained increase in agricultural productivity that gave the world's farmers the ability to produce enough food to feed the entire population of the earth many times over.

And yet mass malnutrition and starvation have persisted to the present day. The question, "Why can't everyone have enough to eat?" is no less urgent today than it was in Malthus's era.

Hunger on so vast a scale suggests that not enough food is being produced to feed the world's people. If that is the problem, then the obvious solution is to increase food production, and that is where agronomy comes in.

The Green Revolution

Based on that commonsense logic, scientists in many countries—but especially the United States—set out to bring to bear the resources of modern science on the curse of world hunger by producing knowledge that could help poor farmers in Asia, Africa, and Latin America grow more food. It began in 1944 with Rockefeller Foundation–funded experiments designed to increase agricultural productivity in Mexico. Their success in raising crop yields was so dramatic that the desire to extend these benefits to the rest of the world was irresistible. That was the debut of the Green Revolution, modern agronomy's answer to global hunger and poverty. By the 1970s, new strains of wheat, rice, and corn, developed by research institutes of the Rockefeller and Ford Foundations, had been propagated throughout the world. The "miracle seeds" of the Green Revolution were accompanied by new farming practices that replaced the traditional methods of millions of poor farmers.

On its own terms, the Green Revolution represented an impressive accomplishment. "The production advances of the Green Revolution are no myth," even its severest critics acknowledge. "Thanks to the new seeds, tens of millions of extra tons of grain a year are being harvested."[3] But there was a catch. Throughout the underdeveloped world, the Green Revolution's growth in food production existed side by side with increased hunger. As paradoxical as that sounds, the explanation is simple: if poor people can't afford to buy food, more food in the market will not help them. The real problem is the vast and growing economic inequality that denies much of the global population access to the necessities of life.

The Green Revolution tended not to meliorate but to exacerbate that inequality. To work their miracle, the new varieties of wheat, rice, and corn required massive inputs of fertilizers and pesticides, which most peasant farmers could not afford. The more affluent growers who could afford them prospered while the poorer farmers had their livelihood destroyed.[4] The big landowners then grew the most profitable cash crops for export rather than staple crops to feed the home population. And the large-scale "farmers" who benefited most of all from the new technology were not in Asia, Africa, or Latin America; they were US-based agribusiness giants such as Monsanto, Cargill, and Archer Daniels Midland.

The Green Revolution accelerated the trend. A 2006 report by UNCTAD, the United Nations Conference on Trade and Development, declared that "concentration in agricultural biotechnology is giving the largest corporations unprecedented power" over food production. The report disclosed that Bayer Crop Science, Syngenta, and BASF had captured half of the agrochemical market, while Monsanto had gained control of one-fifth of the international seed market. Among other things, this has transformed farmers "from 'seed owners' to mere 'licencees' of a patented product."[5]

A key factor driving the distress of small farmers is the immense rise in the prices they have to pay for their seeds. A report entitled *Out of Hand: Farmers Face the Consequences of a Consolidated Seed Industry* describes a single-year increase: "Corn seed prices in 2009 were more than 30 percent higher, and soybean seed nearly 25 percent higher, than 2008 prices."[6]

Since then, the growth and concentration of Big Food has continued apace. By the beginning of 2019, DuPont and Dow had merged, Bayer had acquired Monsanto, and Syngenta was bought by ChemChina, creating three megacompanies that control 61 percent of international seed and pesticide production.

The "appropriation by the large biotech companies of knowledge that was developed in the public domain" intensifies the corporate corruption of agricultural science. The "privatization and patenting of agricultural innovation (gene traits, transformation technologies, and seed germplasm)" has facilitated the agribusiness dominance that has rendered agronomy impotent to solve the great problem it set for itself.[7]

Enter the Gateses and More Free Marketeering

After its impressive period of expansion from 1970 to 1990, the Green Revolution stagnated as the International Monetary Fund and the World Bank pressured the poor countries to buy food on the world market rather than grow their own. In 1999 the Rockefeller Foundation tried to revive the initiative in Africa, but with little effect until 2006, when the Bill and Melinda Gates Foundation rode to the rescue with a $100 million donation to create the Alliance for a Green Revolution in Africa (AGRA).

AGRA's strategy for increasing the food supply—"creative capitalism" with a heavy biotechnology bias—was simply the existing market-based approach with an added emphasis on genetic engineering. Consider this: the market was the medium through which Big Food corporations of the advanced industrial countries had sold farmers of the underdeveloped world seeds, animal feeds, fertilizers, and pesticides, and then bought and processed their raw agricultural products. The upshot was the consolidation of small farms into large agribusinesses and the impoverishment of millions of displaced small-scale farmers. The rich produced more food and the poor got hungrier. In summary, the market cannot solve the problem; it *is* the problem.

The naive belief that the way to end world hunger is to grow more food stems from a profound misreading of the economic history of agriculture. Until a century and a half ago, food production was insufficient to meet the basic needs of the world's population. In that context, Malthus's 1798 prediction that the global food supply would become ever more inadequate did not seem unreasonable. But by the last quarter of the nineteenth century, agricultural productivity had increased to the point that *too much* food was being produced—too much, that is, for the market to absorb. Crop prices plummeted and growing numbers of farmers went bankrupt. A permanent crisis of overproduction in agriculture commenced and has persisted ever since.

Food production is one thing and food production *capacity* quite another. The Industrial Revolution, augmented by the Green Revolution, gave us the capacity to produce more than enough food for everybody, but a great deal of that capacity is not being put to use. In our current market-based system of resource allocation, the decision as to how much food to produce is based primarily not on how many human beings need it but on how much can be profitably sold. Ultimately, food shortages are not what cause world hunger, so producing more food will not end it. The potentially useful science that agronomists practice is wasted because it is beside the point in the context of current socioeconomic reality.

Meanwhile, agriculture's increasing dependence on costly and environmentally destructive chemical fertilizers and pesticides continues to strengthen Big Food's control over the world's food supply. But as their petrochemically dependent production methods began to run up against the law of diminishing returns, the same corporations turned to genetic engineering as another scientific quick fix. "We must be skeptical," critics of the Green Revolution warned at the turn of the twenty-first century, "when Monsanto, DuPont, Novartis, and other chemical-cum-biotechnology companies tell us that genetic engineering will boost crop yields and feed the hungry."[8]

CHAPTER 3

From Green Revolution to Gene Revolution

ARE GENETICALLY MODIFIED ORGANISMS—GMOs for short—necessary to feed the world's growing population? That has been the primary claim advanced on their behalf by their creators and defenders. Their critics respond that the technology's long-term risks to human health and the environment have not been adequately investigated, and that GMOs cannot solve the world hunger problem. The highly polarized debate shows no signs of abating anytime soon.

The specific words "genetic modification" could not have been used until the nineteenth century, when the gene was identified as the biological unit of inheritance, but human beings have been genetically modifying plants and animals for many millennia. The domestication of plants and animals was essentially a matter of manipulating their genetic material; in that sense it is not illegitimate to say that the hunter-gatherers who initiated agriculture were practicing de facto genetic engineering. The vast botanical knowledge they accumulated was a direct antecedent of modern agronomy. There is a crucial distinction to be made, however, between traditional and modern methods of genetic modification.

Traditional versus Transgenic Techniques

Agriculture is not simply a matter of growing plants found in nature and harvesting them; it requires domesticating species of wild plants and transforming them into *crops* designed to serve human needs. In spite of the "natural" labels that proliferate in our supermarkets today, almost none of the foods we eat have ever been truly natural.

Domesticated plants were created neither by nature nor by a providential deity but by prehistoric humans. The original plants were *genetically modified* (GM) in such a way that their modern offspring are unable even to propagate themselves. An ear of corn (maize) is an artifact; it can't exist without humans removing kernels from cobs and planting them. Without human intervention, corn would disappear from the face of the earth in a single generation.

Plant domestication was at first inadvertent, occurring in the course of early humans' daily quest for subsistence. But eventually hunter-gatherers paved the way for cultivation by intentionally experimenting with all the species available in their physical environment. In so doing, they discovered—among hundreds of thousands of wild species of flowering plants—the very few that could be altered to better serve human purposes. In the five centuries since Europeans "discovered" America, not one American plant capable of being domesticated has been identified that the prehistoric Amerindians had not already found, genetically modified, and cultivated.

Selective breeding, cross-hybridization, and grafting are traditional means of producing genetically modified organisms. Modern geneticists sometimes claim they are simply advancing along the path established by those traditional techniques. But transgenic technology is in fact a qualitatively different form of genetic manipulation, one that breaches the natural barriers evolved by organisms as protection against the introduction of DNA from vastly different species.

Traditional methods, for example, could produce a new canine species by mating wolves with coyotes,[1] but the notion of mating a scorpion with a cabbage would have been dismissed as an absurdity. Modern genetic engineers, however, have indeed succeeded in splicing DNA from scorpions into cabbages.[2] Why would anyone want venomous cabbage? Scorpion poison in GM virus form was already being used in pesticides. By engineering it directly into cabbages, less pesticide would theoretically be necessary to keep caterpillars from damaging cabbage crops, with—again, theoretically—no risk to human health.

The Tale of Enviropig™

A pig, a mouse, and an *E. coli* bacterium walk into a bar . . . ? It is no joke that this ménage à trois has actually been consummated by gene splicing to produce a novel creature, trademarked as Enviropig™ by its creators and called "Frankenswine" by public-interest advocates.

Enviropig™ was genetically engineered at the University of Guelph in Canada with a purpose in mind. The alien genes gave it the capacity to digest phosphorous more efficiently, allegedly making it more environmentally friendly, but GMO critics find that claim dubious, disingenuous, and dangerously deceptive.[3] While its creators claim that Enviropig™ was engineered to solve an environmental problem—water pollution caused by runoff from vast pools of hog manure on industrial-scale farms—environmentalists respond that the motivation was in fact economic: to reduce the costs involved in the disposal of millions of gallons of liquefied pig feces.

The critics' point was confirmed when the Enviropig™ project was ended for economic rather than scientific reasons: no company was willing to foot the bill to take the peculiar piggy to market. The genetic modification had been rendered less necessary by the development of a feed supplement to help pigs digest phosphorus. As the price of that supplement decreased, farmers had less incentive to gamble on the porcine oddity.

"Stuff Happens"

Meanwhile, safety and ethical concerns about GM animals mounted. A key worry was that Enviropigs™ might get out of their pens and breed with regular pigs, allowing their modified genes to spread uncontrollably throughout the global porcine population. Then, in 2002, an accident occurred. Eleven Enviropiglets™ were slaughtered, and a bureaucratic snafu allowed their remains to contaminate 675 tons of poultry feed used by egg and turkey farmers. A university research official overseeing the Enviropig™ project said, in a newspaper interview, "Things you don't expect to happen can happen."[4] To which GMO critics everywhere exclaimed: *Our point precisely!*

The consequence to be feared is not that Enviropig™ genes will be transferred into the DNA of animals or humans that ingest them.

Potentially more insidious are the as yet unknown dangers of allow-ing aberrant genes to irreversibly contaminate our food supply—from pigs, via poultry, to us.

GMOs, Pesticides, and Herbicides

Although the direct presence of GMOs in the foods humans eat is what catches the attention of most consumers, in fact the scientific debate over GMOs is more focused on the indirect ways that transgenic technology affects the food chain. It is unlikely that you will ever eat the meat of an Enviropig™, but most of the pork you eat will have been fed on GMO crops. In the United States, 92 percent of the field corn and 94 percent of the soybeans are raised from GMO seed.[5]

Much of the controversy has not concerned GMOs in crops them-selves but the pesticides and herbicides used on crops. Pesticides have been a key environmental concern since the 1960s, when Rachel Carson raised the alarm over the dangers that DDT posed to the world's food supply with her book *Silent Spring*.

The food industry has tried to woo environmentalists with the argu-ment that growing GMO crops leads to a decrease in pesticide use. The evidence, however, does not support that claim. A comprehensive study of the early years of GMO crop experience in the United States—1996 through 2011—reported a net *increase* of four hundred million pounds of pesticide application.[6] Also, comparing American and European agri-culture over the past two decades: herbicide use *increased by 21 percent* in the United States, where GMOs were ubiquitous, while *decreasing by 36 percent* in Europe, where GMOs have largely been rejected.[7]

Unlike herbicides, however, some kinds of pesticides—insecticides and fungicides—did indeed decrease in use significantly after the intro-duction of GMO crops. The reason for that is worth exploring. Consider this: to ward off insects and fungi, crops are genetically modified to repel them. That lessens the need for farmers to use insecticides and fungi-cides. But the way GMO crops avoid weeds is quite different. For that, crops are modified to be resistant to standard herbicides. That allows farmers to spray on *more* herbicides without damaging their crops.

The logic is impeccable. By providing herbicide-resistant seeds that produce crops needing more herbicide, the agrochemical firms that sell both seeds and herbicides make a double killing. By contrast, selling insect-resistant seeds undercuts the market for their insecticides. Monsanto's projected next generation of GMO corn will be engineered to resist five different weed killers. The goal, as one observer noted, is always "to sell more product."[8]

The Battle Lines in the GMO Debate

The impassioned dispute over GMOs is a legitimate scientific debate, with competent scientists on both sides. Two basic questions predominate:

> Are GMOs necessary?
> Are GMOs safe or do they pose a grave environmental danger?

These issues abound with strong political and economic ramifications. A related concern is thoroughly political: Should the food industry be required by law to label its products so consumers can know whether or not they contain GMOs?

Some proponents of GMOs have tried to marginalize their critics by lumping them together with creationists and climate-change deniers. That false imputation has gained some traction, as the March 2015 *National Geographic* cover demonstrated by listing five examples of "anti-science": "Climate change does not exist," "Evolution never happened," "The moon landing was fake," "Vaccinations can lead to autism," and "Genetically modified food is evil." The accompanying article paid scant attention to genetic engineering per se. Unfortunately, Neil deGrasse Tyson, reputed to be "perhaps the most credible public scientist on the planet," has also lent his prestige to disparaging GMO critics.[9]

The dismissal of GMO opponents as fringe elements echoes the notion that a broad scientific consensus on the issue has been achieved. That claim has been disputed by a statement signed by three hundred dissenting scientists and scholars, bluntly declaring: "There is no scientific consensus on the safety of genetically modified foods and crops."[10]

Public Counterweight to Corporate Influence?

The appearance of pro-GMO consensus is bolstered by the official positions of the premier organizations of American scientists—the National Academies of Science, Engineering and Medicine, and the American

Association for the Advancement of Science. There can be little doubt that most professional scientists in the United States do indeed support the use of transgenic technology.

The National Academies are expected to function as a public counter-weight to private interests, but as establishment US institutions they are far from immune to economic and political pressures. The ways Big Food manipulates the pillars of public science are usually hidden from view, but they were brought to light by the exposure of conflicts of interest at the heart of these nominally independent institutions of official science. A major report by the National Academies of Sciences, Engineering and Medicine that vouched for the safety of GMOs[11] was shown to have been written by a committee that included several scientists with undisclosed ties to the biotech industry. Of the panel's twenty members, six had one or more financial conflicts of interest.[12]

That panel's recommendations, a *New York Times* investigation noted,

> could have broad implications for the industry. Big food compa-nies like Coca-Cola and Archer Daniels Midland, for instance, have invested in synthetic biology, a term used for the more so-phisticated genetic engineering now coming into use that was the panel's area of study. Food companies are exploring its use in cre-ating flavorings and sweeteners.[13]

The most blatant conflict of interest involved one Douglas Fried-man, the person responsible for forming the panel of scientists that is-sued the academies' report. While Friedman was choosing its members, he was also applying for a job in the biotech industry. Of the thirteen scientists he selected, five were directors or advisors of his employer to be, Engineering Biology Research Consortium. The panel was formed in April, and three months later Friedman was Engineering Biology Re-search Consortium's new executive director. When asked to explain this, "he said it was a coincidence."[14] William Kearney, a spokesman for the academies, acknowledged that two scientists were named to the panel despite known conflicts of interest. It was impossible, he said, to find people with the requisite biotech expertise who were not in some way connected to the industry.

That, in microcosm, explains the essential dilemma of American science: in a society unwilling to adequately fund independent research, the men and women who generate and vet scientific knowledge—whether in university, government, or corporate settings—must ultimately depend on private sources to earn a living. Knowledge produced under those conditions cannot be assumed to be objective. *Who pays the piper calls the tune.*

How the Pipers Are Paid

Behind-the-scenes funding by a food industry heavily invested in GMOs undermines the integrity of the public discourse. The scientists who serve as spokespersons for the technology are consciously or unwittingly fronting for Bayer-Monsanto, DowDuPont, and other powerful corporate interests that bankroll their research and their careers.

Big Food puts these scientists forward as the human face of their permanent PR campaign on behalf of GMOs. Its trade associations sponsor front groups and websites posing as independent sources of scientific information, often the same ones it created to mold nutrition science to its purposes. The cases of two shills for Monsanto, Dr. Kevin Folta and Dr. David R. Shaw, will be presented below. These activities normally proceed under cover of secrecy, but the public exposure of an industry trade group's internal emails reveal in detail how prominent scientists are recruited and groomed to pose as independent endorsers of its member firms' products.

The trade group was the Council for Biotechnology Information, which euphemistically identified itself as a "coalition of life-science companies," but its constituents happened to include the Big Six manufacturers of genetically modified seeds.[15] In 2013 the council created a website called GMOAnswers.com "to provide consumers with balanced, fact-based responses to their questions." The website was developed by a public relations firm, Ketchum Inc., known for serving "corporate clients fighting environmental organizations opposed to their products or actions."[16]

GMOAnswers.com says it offers responses from "independent experts" who "are not affiliated with the Council for Biotechnology Information or its member companies." But a public-interest advocacy group named US Right to Know challenged those claims by filing requests under the Freedom of Information Act.[17]

"The Gloss of Impartiality and Weight of Authority"

As a result, thousands of pages of emails were released that describe in detail how Monsanto and other food industry giants developed a "public relations strategy to spotlight a rarefied group of advocates: academics, brought in for the *gloss of impartiality and weight of authority* that come with a professor's pedigree." The emails "show how academics have shifted from researchers to actors in lobbying and corporate public relations campaigns." They reveal that Big Food not only "aggressively recruited academic researchers" but also that "the biotech industry has published dozens of articles, under the names of prominent academics, that in some cases were drafted by industry consultants."[18]

Those consultants were from Ketchum, the PR firm. One of the scientists they recruited to provide the answers for GMOAnswers.com was Kevin Folta, chair of the Horticultural Sciences Department at the University of Florida. Folta "is among the most aggressive and prolific biotech proponents, although until his emails were released," he "had not publicly acknowledged the extent of his ties to Monsanto." But the PR firm "did more than provide questions. On several occasions, it also gave Dr. Folta draft answers, which he then used nearly verbatim." In August 2014, Monsanto gave Dr. Folta "a grant for $25,000 to allow him to travel more extensively to give talks on the genetically modified food industry's products." After receiving the grant, Dr. Folta sent a thank-you email to a Monsanto executive, saying, "I am grateful for this opportunity and promise a solid return on the investment."[19]

Molding public opinion is not the only service scientists-for-hire provide. As lobbyists they have "helped produce important payoffs, including the approval by federal regulators of new genetically modified seeds after academic experts intervened with the United States Department of Agriculture on the industry's behalf." Dr. Folta, for example,

> became part of an inner circle of industry consultants, lobbyists and executives who devised strategy on how to block state efforts to mandate G.M.O. labeling and, most recently, on how to get Congress to pass legislation that would pre-empt any state from taking such a step.[20]

Another university scientist the agribusiness giants enlisted is David R. Shaw, vice president for research and economic development at Mississippi State University. Monsanto has supported Dr. Shaw to the tune of at least $880,000 in grants to fund the research projects he oversees. Dow Chemical also sought Dr. Shaw's help in soliciting the Department of Agriculture's approval of a new genetically modified cottonseed, reminding him in an email "about the industry's financial support for the university."[21]

Are GMOs Necessary?

Proponents of genetic modification technology argue that it is necessary to feed a hungry world. The critics' response is logically neat and simple: if the problem is not insufficient food production, then increasing food production capacity, which is what GMOs offer, cannot be a solution. QED.

The validity of that syllogism depends on the premise. The previous discussion of the Green Revolution concluded that increasing food production capacity is a technocratic strategy that ignores social reality. The root problem of hunger is economic inequality. The billion or so hungry people who can't afford to buy enough food will remain hungry no matter how much food is available on the market.

The most influential scientific support for GMOs comes from the National Academies of Science, Engineering and Medicine. Its major report on the subject was produced, as noted above, by a committee with significant ties to the biotech industry. The report, not surprisingly, leaned heavily to the technocratic side. Nonetheless, it acknowledged the technology's limits: "Even if a GE crop may improve productivity or nutritional quality, its ability to benefit intended stake-holders will depend on the social and economic contexts in which the technology is developed and diffused."[22]

But Wait! Do GMO Crops Even Produce More Food?

Let's back up for a moment. Aside from the question whether GMOs are necessary to feed a hungry world, do they even succeed in their most basic promise—to increase crop yields? The evidence points in the opposite direction. Even the aforementioned National Academy of Science's 2016 report supporting transgenic technology found "no evidence from USDA data that the average historical rate of increase in U.S. yields of cotton, maize, and soybean has changed" since the adoption of GMO agriculture.[23]

Furthermore, an in-depth analysis by the *New York Times* arrived at the same conclusion after comparing two decades of crop yields in Europe—where GM technology has been sharply restricted by strong regulations—with those in North America, where it has been used extensively. "United Nations data showed that the United States and Canada have gained no discernible advantage in yields—food per acre—when measured against Western Europe."[24]

The *Times* investigation focused on three major crops: rapeseed (the source of canola oil), corn (maize), and sugar beets. For rapeseed, the Canadian output per acre was compared to Western Europe's. "Despite rejecting genetically modified crops, Western Europe maintained a lead over Canada in yields."[25] For corn and sugar beets, the comparisons were between the United States and Western Europe. With corn, "the trend lines between the two barely deviate." And sugar beets "have shown stronger yield growth recently in Western Europe than the United States, despite the dominance of genetically modified varieties over the last decade."[26]

The American food industry's claim that its GM technology is indispensable to feeding a hungry world is apparently hollow.

Are GMOs Safe?

The GMO safety issue could provide material for a long series of *New Yorker* cartoons. Picture two geneticists in white lab coats holding test tubes and syringes. Sample captions might be:

> "Let's see what happens if we put a scorpion's poison-producing genes into a cabbage plant. Maybe we'll get a cabbage with heightened resistance to pesticides. *What could go wrong?*"

> Or: "How about we splice an anthrax gene into an *E. coli* bacterium? Maybe we'll get an antidote for anthrax poisoning. *What could go wrong?*"

GMO safety, however, is no laughing matter. Critics of GM technology have expressed the fear that altered genes entering the biosphere will do irreversible damage. It is undeniable that transgenic organisms have already contaminated food chains by jumping fences into adjacent fields of crops and crossbreeding with populations of their wild relatives. A database of recorded cases of GMO contamination listed 396 incidents in 63 countries from 1997 through 2013.[27]

Meanwhile, the GMOs' defenders contend that current widespread GMO consumption without serious incident is sufficient evidence of their safety. That argument, the critics point out, is analogous to the one nuclear power advocates used before the Chernobyl and Fukushima disasters.[28] The appeal to scientific evidence has been one-sided. GMO proponents point to thousands upon thousands of studies indicating that transgenic technology poses no dangers to human health,[29] while GMO critics offer no "smoking gun" proof in rebuttal. That, the critics insist, is beside the point, because without "credible, independent, long-term" studies, "the safety of GMOs is *unknown*,"[30] and the burden of proof must lie with those who claim otherwise. Thus far, the mountain of research supporting the safety of GMOs does not meet the criterion of independence because it has been unduly influenced by powerful agribusiness and biochemical corporations with vested interests.

Glyphosate, Monsanto, Roundup, and GMO Safety

Glyphosate, "the most heavily used agricultural chemical in history," is at the heart of the dispute over the safety of genetically engineered crops. It is the active ingredient in Monsanto's popular weed killer, brand-named Roundup.[31] As investigative journalist Carey Gillam explains:

> From the day genetically engineered crops were introduced, they were designed with one primary purpose in mind—to withstand treatments of glyphosate. . . . Then and now, most of the genetically modified crops grown in the world carry the glyphosate-tolerant trait.[32]

That genetically engineered "glyphosate-tolerant trait" is the key to understanding how Monsanto came to dominate both the global markets for agricultural chemicals and for crop seeds. By contractually linking its weed killer to its seeds, Monsanto gained an enduring pipeline of multibillion-dollar profits from combined sales of seeds and herbicide.

Monsanto promoted Roundup as much safer for human health than all previous herbicides. Unfortunately, its claims proved to be untrue. The brilliant technological gambit of utilizing transgenic techniques to create glyphosate-immune crops has a fatal flaw: it kills people. In 2015, a prestigious United Nations–affiliated health agency, the International Agency for Research on Cancer (IARC), designated glyphosate as "a probable carcinogen in humans," and "found a particular association between glyphosate-based pesticides and non-Hodgkin lymphoma."[33]

As a consequence, Monsanto was confronted with more than 130,000 lawsuits from plaintiffs charging that its weed killers had caused them to develop non-Hodgkin lymphoma (NHL) and other deadly cancers.[34] In 2020, Bayer (which in 2018 had assumed Monsanto's obligations due to a merger) agreed to pay between $10.1 and $10.9 billion to settle almost a hundred thousand of those claims.

Bayer continued to insist that no scientific evidence proves that its herbicides are carcinogenic. Nonetheless, that massive multibillion-dollar payout was a tacit admission of glyphosate's culpability in causing NHL, which completely undermines the American scientific establishment's longstanding refusal to confront the issue of GMO safety. The evidence of the danger that Roundup poses to human health, and of the corruption of science by corporate influence, solidifies the case that oversight of agrochemicals and GMOs alike must be governed by the precautionary principle.

The Precautionary Principle

The American Academy of Environmental Medicine warns that "GM foods pose a serious health risk in the areas of toxicology, allergy and immune function, reproductive health, and metabolic, physiologic and genetic health." For those reasons, "the AAEM believes that it is imperative to adopt the precautionary principle."[35]

The precautionary principle is an ethical and legal norm that guides the European Union's GMO regulatory policies. Its essence is that waiting for scientific certainty of negative consequences before taking action to protect human health and the environment is unconscionable. US regulatory agencies explicitly reject the precautionary principle in favor of a rival "decision-making model" that goes by the name Risk Assessment.[36] The EPA defines Risk Assessment as a tool to objectively determine "the nature and magnitude of health risks to humans . . . from chemical contaminants and other stressors that may be present in the environment."[37] That is certainly a reasonable aspiration, but in practice

Risk Assessment is based on shadowy science manipulated by the corporations that the agencies are supposed to regulate.

Public-interest advocate Carol Dansereau has eloquently described how "the precautionary principle gave way to a policy of gambling with health and the environment" in the United States. Dansereau undertook a search for the "sound science" the regulatory agencies claim underlies their Risk Assessment model and found only "hidden, unpublished corporate-funded studies from decades ago."[38] Dansereau also explains how agribusiness interests use patent laws to prevent independent investigations of the safety of their patented GMOs. She quotes a 2009 protest to the EPA by twenty-six university scientists from sixteen states, stating that "as a result of restricted access, no truly independent research can be legally conducted on many critical questions."

Considering all that is unknown about the long-term effects of transgenic technology, is it not eminently reasonable to err on the side of caution? Unfortunately, American science policymakers have chosen to err on the side of profits. GMOs have been present in American food for two decades, and now are in more than 80 percent of the processed foods we eat. "Most packaged foods contain ingredients derived from corn, soy, canola, and sugar beet—and the vast majority of those crops grown in North America are genetically modified."[39]

Should GMOs Be Labeled?

A major theme of the GMO debate has been whether the law should require foods containing genetically engineered ingredients to be labeled to say so. England, France, Germany, Japan, Australia, and dozens of other countries around the world passed such laws in the first decade of the twenty-first century, but the United States and Canada were not among them.

This was a political rather than a scientific dispute, and the battle lines were clearly drawn. A 2015 ABC News poll found that 93 percent of Americans favored GMO labeling. A grass roots "right to know" movement arose to press that demand, and campaigns for labeling laws gained traction. Realizing that any products with "Contains GMOs" labels would be at a great disadvantage in the supermarkets, Big Food pushed back, spending more than $103 million between 2012 and 2014 on PR campaigns that successfully beat back label-law initiatives in a number of states.[40]

The conflict came to a head in 2016 when Vermont passed a law requiring GMO labeling. The food industry responded by shifting their lobbying efforts into high gear behind a Senate vote on a federal law to rescind all such state laws. Public-interest advocates dubbed this the DARK ("Deny Americans the Right to Know") Act. The public support for GMO labeling proved to be too overwhelming to be ignored. In May 2016 the DARK Act went down to defeat.

The agribusiness interests regrouped and tried a new tack. They lobbied for "DARK Act II": a federal law mandating GMO labeling but allowing the corporations to make the notification in less direct ways. One would be by placing a QR ("quick response") code on the label that a consumer could read with their smartphone. Another would put a 1-800 number or website address on the label that would direct the consumer to information about GMO ingredients.

The ploy was laughably obvious. Millions of Americans who don't own smartphones cannot read QR codes, and the inconvenience of accessing websites or call-in numbers while grocery shopping would deter almost everyone from using them. The food industry could thereby continue to hide what it wants to hide.

Nonetheless, DARK Act II passed with bipartisan support, and on July 29, 2016, Barack Obama signed it into law. Both he and the food industry lauded it as a big step forward for consumer rights. Consumer advocates blasted it as a sham.

A genuine GMO label law would have been a major victory for American consumers, but it would not have resolved the GMO controversy. It could conceivably have forced the food industry to stop using GMO ingredients simply by destroying the market for foods containing them, but the scientific issues remain in contention.

A Final Word on Science, Anti-science, and GMOs

The vast majority of consumers will continue to be wary of GMO ingredients in their food, and the majority of American scientists may continue to disparage those consumers as irrational and "anti-science." In June 2016 a well-publicized open letter signed by more than a hundred Nobel Prize laureates charged that "organizations opposed to modern plant breeding, with Greenpeace at their head," have misrepresented the "risks, benefits, and impacts" of GMOs "and supported the criminal destruction of approved field trials and research projects."[41] Greenpeace has employed civil disobedience and direct action tactics, including symbolic destruction of GMO field trials, in its protest activities, but the activists are no more motivated by criminality than the Plowshares peace protestors who have repeatedly broken into weapons plants and "vandalized government property."[42]

While many of the individual Nobel laureates may not have ties to the food industry, the open letter was by no means an expression of independent science. It was organized by a mysterious new group called Support Precision Agriculture, which to no one's surprise turned out to be a front created by biotech PR firms.[43]

The letter's principal argument is that in opposing the GM development of "Golden Rice," Greenpeace and other environmentalists are blocking an effort to "eliminate much of the death and disease caused by a vitamin A deficiency (VAD), which has the greatest impact on the poorest people in Africa and Southeast Asia." It concludes by provocatively asking, "How many poor people in the world must die before we consider this a 'crime against humanity'?"[44]

Vitamin A deficiency does indeed cause blindness in hundreds of thousands of African and Asian children every year. But a *New York Times* exposure of "G.M.O. Misconceptions" included a critique of the claim that Golden Rice "is saving the lives of malnourished children in the developing world." In fact, it has been "in development" for two decades and shows little promise of ever actually being produced. In the spring of 2016, "the nonprofit institute responsible for the rice's development said it had gone back to the drawing board because the rice strains that had the improved nutrition did not grow well enough to be embraced by farmers."[45]

The Nobel laureates who signed the letter would do well to reflect on their naivete in endorsing a scientific position without examining its premises. They would no doubt be shocked by the evidence that GM technology—despite food industry claims—has, in many instances, not been able to produce more food per acre than non-GMO agriculture.[46]

Are American consumers who eschew GMOs really irrational and "anti-science"? I would argue that they are fully rational on two counts: on the basis of the precautionary principle, and in their distrust of experts who are on the payroll of the food industry. As for truly independent scientists who condemn critics of GM technology, they are unwittingly doing the public a great disservice.

CHAPTER 4

The Tobacco Strategy

THE HEADLINE READ: "NOW... Scientific Evidence on Effects of Smoking!" And what did the data show? That smoking Chesterfields produced "no adverse effects on the nose, throat, and sinuses."[1] How quaint and bittersweet-humorous those cigarette ads of yesteryear seem today.

But back to the real world. Since the advent of mass-produced cigarettes, lung cancer has risen from being a rare condition representing only 1 percent of all cancers to its present status as the number one cause of cancer deaths for both men and women in the United States. It currently kills 1,760,000 people a year worldwide, including about 140,000 Americans.[2] Meanwhile, independent medical science has demonstrated the link between smoking and lung cancer far beyond reasonable doubt. Further research into the etiologies of emphysema, asthma, bronchitis, heart disease, and numerous other serious maladies laid bare the shocking extent of the damage to human health the cigarette makers' products have caused over the years.

The tobacco industry, rather than recognizing its social responsibility to reorient its agricultural and manufacturing resources to create nonhazardous commodities, instead chose to double down by attempting to discredit the science. Its efforts bore poisonous fruit.

Big Tobacco, False Doubt, and False Equivalence
Big Tobacco's efforts to corrupt the health sciences have been analyzed by historians of science Naomi Oreskes and Erik Conway in a book entitled *Merchants of Doubt*.[3] Its considerable impact on the public discourse was

further amplified by the documentary film it inspired. *Merchants of Doubt*'s history of the tobacco industry's activities provides an indispensable basis for understanding the corporate efforts to debase science in general. The main ploys of the "tobacco strategy" were to inject doubt about legitimate scientific conclusions into the national conversation, to thereby manufacture public controversies that did not exist among scientists, and to demand equal time in the mass media for "both sides" of the contrived disputes.

A remarkable self-exposure of this strategy was discovered among the internal communications of the Philip Morris Company: a 254-page how-to manual entitled *Bad Science: A Resource Book*. It provides ammunition to battle regulatory agencies like the EPA that "allow political objectives to guide scientific research" and "enable bad science to become a poor excuse for enacting new laws and jeopardizing individual liberties."[4] Unfortunately, the stratagems it teaches have worked all too well. Big Tobacco's later imitators followed its lead by challenging independent research into the effects on human health of acid rain, pesticides, and the hole in the ozone layer, to mention but a few.

How the Doubt Industry Began

Scientific evidence of tobacco's carcinogenic and addictive properties began to surface in the 1950s. Large-scale tobacco growers and cigarette manufacturers recognized the myriad threats to their industry that the revelations posed. First, they feared a sharp decline in the market for their products. Second, they worried that legislatures and regulatory agencies might hamstring their profitable operations. And third, they anticipated an onslaught of crippling litigation holding them accountable for the hundreds of thousands of deaths their products caused every year. The birth of "product defense science" was a major milestone in the corruption of American science.

The leading cigarette manufacturers—Philip Morris, R. J. Reynolds, Benson and Hedges, and others—joined hands and hired a PR firm, Hill & Knowlton, to defend them in the court of public opinion. The tobacco moguls were advised to fight independent science with science they could control, which they did with great success.

John Hill of Hill & Knowlton told the tobacco executives that it would be futile to outright deny the medical research. As an alternative, he advised that they hire reputable scientists to produce genuine findings about other possible causes of cancer, heart disease, or any malady that cigarettes had been blamed for. That way, when regulators or litigators attempted to assign culpability to cigarettes, the industry could respond with what they called "sound science" to back it up—that smoking *could not be said with absolute certainty* to have caused the disorders. The benign phrase "sound science" was thus transformed into Orwellian doublespeak signifying "industry-funded smokescreen masquerading as science."

Hill insisted that the public's trust could not be gained "through advertising, which was self-interested by definition." Instead, "it would be crucial for the industry to assert its authority over the scientific domain; science had the distinct advantage of its reputation for disinterestedness."[5]

Enter the Physicists

Foremost among the scientists who enlisted in the cause of Big Tobacco were two retired physicists named Fred. Frederick Seitz, an atomic physicist, and S. Fred Singer, a rocket scientist, had made names for themselves in the Manhattan Project, which created the atomic bomb during the Second World War. The two Freds set up a front group, the Council for Tobacco Research, to serve as the public face of pro-tobacco science. Their endeavor was generously endowed. By the end of the 1970s, the industry had

> spent over $50 million on biomedical research. Individual tobacco companies [had] invested millions more—bringing the total to over $70 million. By the mid-1980s, that figure had exceeded $100 million.[6]

Seitz and Singer were later joined by two more physicists, William Nierenberg and Robert Jastrow, in founding a think tank they named the George C. Marshall Institute. The Marshall Institute served as the vehicle for transforming the tobacco strategy into a general crusade against independent science. The four men's military-industrial and Cold War roots were by no means irrelevant to their cause; their shared outlook included anticommunism and Ayn-Randian laissez-faire, individualist ideology.

A common theme of their campaigns on behalf of Big Tobacco, Big Coal, Big Oil, and the military-industrial complex was the proposition that government oversight of industry is a slippery slope to communism. Despite the demise of the Soviet Union, the quartet of Cold Warriors continued their offensive against genuine science through the 1990s and into the twenty-first century. Their efforts would have gone nowhere, however, without the funding provided to them by the tobacco, fossil fuel, and weapons industries.

The Triumph of Doubt and Disinformation

Thanks to the largesse of the tobacco industry's lobbyists, politicians needed only the slightest whiff of science to be persuaded not to burden the cigarette manufacturers with onerous regulations. And for decades, all of the many lawsuits against Big Tobacco were stymied by corporate defense attorneys and their scientific experts, skillfully arguing reasonable doubt.

Public opinion was, as intended, confused. The PR campaign not only hired highly credentialed scientists to front for the industry but also aggressively promoted the public perception of a raging scientific debate where there was none. The only scientists who denied that smoking causes lung cancer and heart disease were the ones the tobacco companies paid to do so. A major aim of the PR strategy was to browbeat the mass media into recognizing the manufactured scientific debate as a legitimate one. The Fairness Doctrine required that broadcasters give equal time to both sides of contentious issues.[7] Although the denial side of the "smoking kills" controversy was bogus, its backers—while hiding where their funding came from—mobilized sufficient material resources and political connections to persuade journalists of its authenticity. As a result, a false equivalence was established in the public discourse to "balance" paid-for PR against legitimate science.

The public relations campaigns began with a one-two punch. The first blow, on January 5, 1954, was a two-page newspaper advertisement headlined "A Frank Statement to Cigarette Smokers" that ran in 448 papers with a total readership of forty-three million. Three months later, a booklet entitled "A Scientific Perspective on the Cigarette Controversy" was issued that cited the doubts of thirty-six scientists about cigarettes'

role in causing diseases. A confidential internal memo from Hill &
Knowlton, which produced the booklet, says it

> was sent to 176,800 doctors, general practitioners and special-
> ists, . . . deans of medical and dental colleges . . . [and] 114 key
> publishers and media heads. . . . Days in advance, key press, net-
> work, wire services and columnist contacts were alerted by phone
> and in person . . . in Los Angeles, Chicago, Cleveland, Pittsburgh
> and Washington, D.C. The story was carried by hundreds of pa-
> pers and radio stations throughout the country. . . . Staff-written
> stories [were] developed with the help of Hill & Knowlton, Inc.
> field offices.[8]

Seitz and Singer's disinformation operations began with doubt and dis-
traction but eventually devolved into reprehensible personal attacks against
independent scientists. Unfortunately, this has become the standard modus
operandi of the anti-science campaigns. Overt hate speech in the name of
opposing "political correctness" did not begin with Donald Trump.

There Can Be No Doubt about the Merchants of Doubt

The assertion that Big Tobacco used science to deliberately mislead the
public on this life-and-death issue is confirmed by the documents of the
tobacco companies themselves.[9] The industry's own internal communications prove that its researchers knew numerous deadly truths—that smoking kills, that secondhand smoke kills, and that nicotine is addictive—at
least by the early 1960s but nonetheless persisted in sowing doubt. Its success in fending off regulation and lawsuits for several decades allowed it to
continue reaping billions of dollars in profits from its poisonous commerce.

Anonymous whistleblowers within tobacco corporations secretly
copied thousands of pages of internal documents and leaked them to
public-interest advocates and the press. Those disclosures were later
dwarfed by *eighty million* pages made public as a result of litigation. For
one iconic example of their contents, this line from a 1969 memo circulated internally at the Brown and Williamson corporation gives the
whole pseudoscientific game away: "*Doubt is our product* since it is the
best means of competing with the 'body of fact' that exists in the mind
of the general public. It is also the means of establishing a controversy."[10]

Ironically, making the massive archives public had an unintended negative consequence. While exposing the tobacco corporations' malfeasance, they also provided a detailed blueprint for other industries to follow, thus proliferating the tobacco strategy of discrediting genuine science by means of "situational science." The fossil fuel industry was especially attentive.

And at the End of the Day . . .

Big Tobacco's manufacture of doubt about the etiology of serious diseases succeeded for decades, but its defenses eventually began to crumble. Lawsuits finally succeeded in holding them accountable. Surgeon Generals' reports began officially warning Americans not to smoke. And although regulatory agencies did not ban cigarettes outright, at least they were heavily taxed, disallowed in many public and private spaces, required to bear warning labels, and prohibited from being advertised in key media.

As encouraging as these victories for public health may seem, they are meager in the face of mortality statistics telling us that, *still today,* one in five deaths in the United States—480,000 per year—are caused by cigarette smoking.[11] The global annual death toll is eight million and rising.[12] Consider this: in a typical year, 1981, the tobacco industry spent more than twenty times as much on research as did the American Cancer Society and American Lung Association combined.[13] The industry outspent the legitimate researchers $6,300,000 to $300,000. Imagine what medical science might have accomplished if, year in and year out, those resources had been devoted to trying to cure diseases rather than disguising their causes.

CHAPTER 5

Fraudulent Pharma

MAJOR DRUG MANUFACTURERS such as Pfizer and Merck are, like Philip Morris and R. J. Reynolds, gigantic corporations with billions of dollars in annual sales. While the tobacco industry's abuse of science is total and has only harmful consequences, the pharmaceutical industry's research, by contrast, often produces valid, therapeutically valuable results. Like the medicines it manufactures, however, the benefits of Big Pharma's scientific practices come at a high price and can have dangerous side effects.

"Superstars of Profitability"

Big Pharma's business model rests upon sustaining the perception—sometimes but not always grounded in reality—that its products are of life-and-death importance to those who buy them. The industry has exploited that advantage to gain a powerful grip on the public imagination and the public pocketbook.

Year in and year out, Big Pharma is at or near the top of the list of most profitable industries. For 2015, *Forbes* magazine ranked Health Technology number one among nineteen major American industry groups, with a 21 percent net profit margin. Furthermore, says *Forbes*, "within the broader healthcare technology category, the superstars of profitability go to major and generic pharmaceutical companies, such as Pfizer (27.6 percent), Merck & Co. (25.2 percent), Johnson & Johnson (24.5 percent)."[1]

The Opioids Catastrophe

The most catastrophic of Big Pharma's transgressions in the first two decades of the twenty-first century was the nationwide epidemic of opioid addiction and overdoses it instigated. Traditional villains like opium and heroin were eclipsed by their semisynthetic cousins, oxycodone and hydrocodone, as the prescription addiction crisis spread from often-neglected fringes of American society to the urban, suburban, and rural heartland.

The dimensions of this epidemic are astonishing: more than four hundred thousand opioid-related deaths have occurred in the United States since 2000, according to the US Centers for Disease Control and Prevention. "In recent years, opioid overdoses have been the nation's largest cause of accidental deaths, ahead of even automobile accidents."[2]

A database maintained by the US Drug Enforcement Administration, hidden from public view until 2019, when investigative journalists forced it into the open, provided conclusive evidence of the industry's responsibility for the crisis. "America's largest drug companies saturated the country with 76 billion oxycodone and hydrocodone pain pills from 2006 through 2012 as the nation's deadliest drug epidemic spun out of control."[3]

American medical science facilitated the epidemic. It did so via the corruption of a promising new field of scientific inquiry, pain medicine. Had pain medicine research not been manipulated by an industry seeking to profit from pain by peddling dangerous drugs, it might have yielded great benefits to the many people who suffer from chronic pain.

In the 1980s, the American Pain Society was an organization of scientists and clinicians that promoted a broad spectrum of treatments for pain and was cautious with regard to opioid use. But as the APS's reliance upon pharmaceutical funding deepened, it morphed into an advocacy group—"cheerleaders for opioids," as one US senator described it.[4] By 1996 the APS was proclaiming opioids to be safe and effective for pain management and denying that addiction posed a serious risk.

"They're Getting Away with Murder"

But by 2019, the danger and damage had become so apparent that the major opioid manufacturers and distributers were facing lawsuits from forty-eight states and more than two thousand local and tribal governments. A primary target was Purdue Pharma, which, although it accounted for only about 3 percent of opioid production, was widely perceived as having triggered the epidemic by aggressively marketing its painkiller OxyContin. Another was Johnson & Johnson.

On August 26, 2019, Big Pharma's legal defenses began to disintegrate when an Oklahoma district court ruled that Johnson & Johnson's "false, misleading, and dangerous marketing campaigns" had "caused exponentially increasing rates" of addiction and overdose deaths.[5] Although the ruling directly applied only to Johnson & Johnson, it struck fear into the heart of the industry as a whole.

Within days, Purdue Pharma's owners, the multibillionaire Sackler family, made an offer to settle the thousands of state and federal lawsuits brought against them. Trying to cut their losses, the Sacklers tentatively agreed to give up their ownership of Purdue and disburse between $7 billion and $9 billion from the company plus $3 billion of their own money to the towns, cities, states, and tribes that were suing them.[6]

Most of the plaintiffs indicated willingness to take the deal, but the effort failed when many of the states' attorneys general denounced it as inadequate. Suspecting that the Sackler family was hiding much of the wealth it had accumulated from opioids profits, the New York State attorney general "sent subpoenas to 33 financial institutions and investment advisors tied to the Sackler family as they seek to trace billions of dollars that prosecutors believe the family took out of Purdue."[7]

The lawsuits targeted not only manufacturers like Purdue but distributors of the deadly drugs as well, and the three largest also sought to settle out of court.[8] They and two other defendants offered a deal worth almost $50 billion, but it, too, was rejected. One critic of the proposal declared, "$50 billion is pocket change to them. We need some accountability and it needs to lead to criminal charges. They're getting away with murder."[9]

Market-Driven Distortion of Pharmaceutical Science

The opioid epidemic was Big Pharma's most spectacular crime, but its routine, day-to-day activities are no less contrary to the legitimate practice of pharmaceutical science. The industry's great profitability ultimately derives from its use of science to produce effective medicines. Its primary aim, however, is to produce *marketable* drugs. If they happen to have therapeutic value, all the better, but their value as commodities is what drives corporate R&D.

If research priorities were guided by concern for human health needs, their direction would be dramatically different. The market orientation of the pharmaceutical industry distorts those priorities in numerous ways. R&D dollars flow toward drugs according to their potential market size, which means that more will be spent on research into erectile dysfunction or male pattern baldness than on some of the conditions that pose more fundamental threats to human well-being. By way of example, in 2008 Pfizer announced a major shift in its research plans. Its projected $7.5 billion annual R&D budget was to refocus on "diseases that have high potential for big profits." Six priority areas were identified: cancer, pain, inflammation, diabetes, Alzheimer's disease, and schizophrenia. All of these are certainly worthy of the utmost scientific attention. At the same time, however, Pfizer intended to end its research into "anemia, bone health, gastrointestinal disorders, muscle diseases, obesity, some approaches to osteoarthritis," and "four areas of cardiac disease: hardening of the arteries, high cholesterol, heart failure and peripheral arterial disease."[10]

If those decisions happened to align with the optimal use of resources from the standpoint of human needs, it would have been coincidental. As a financial analyst explained, Pfizer was "basically looking at the largest-margin, largest-market indications." Alzheimer's, for example, "will be a huge market, given the aging population."[11]

Orphan Diseases

An "orphan disease" is defined by the Food and Drug Administration as

> a condition that affects fewer than 200,000 people nationwide. This includes diseases as familiar as cystic fibrosis, Lou Gehrig's disease, and Tourette's syndrome, and as unfamiliar as Hamburger disease, Job syndrome, and acromegaly, or "gigantism." Some diseases have patient populations of fewer than a hundred. Collectively, however, they affect as many as 25 million Americans.[12]

The funding neglect of orphan diseases should prompt governmental agencies to fill the gap, but it has not. Instead, billionaires have begun privately funding medical research into diseases that personally affect them or their loved ones. For now, it still amounts to only a small fraction of overall science spending in the United States, but this "richly financed private research" signals "a profound change taking place in the way science is paid for and practiced in America."[13]

This is not a change for the better. It may appeal to people with romantic notions of swashbuckling financiers and maverick scientists, but when individuals can set private research priorities that affect entire societies, it erodes the ideal of science for the common good. One specific problem is that because the philanthropists' medical priorities derive from their own family histories, their funding is biased toward finding cures for diseases that primarily affect white people, like melanoma, cystic fibrosis, and ovarian cancer. Sickle-cell anemia, by contrast, "mainly strikes black people and has long been something of a research orphan," despite having "crippled and killed millions of people."[14]

Orphan Therapies

In addition to orphan diseases, there are also "orphan therapies" or "financial orphans." These are medicines and protocols with high therapeutic potential but low commercial potential. Although they may show promise as treatments for high-profile diseases, the drug corporations will not spend money to research them because they do not offer enough financial reward. Without the expensive large-scale clinical trials that the FDA rightfully requires to demonstrate a medicine's safety and efficacy, the orphan therapies will remain untested in humans, and unused.

The lower-cost alternatives are shunned despite the fact that American taxpayers partially subsidize the pharmaceutical research. As Dr. Michelle Holmes, who teaches medicine at Harvard, puts it, "What is scientific and sexy is driven by what can be monetized."[15] Dr. Holmes speaks from personal experience. Her research has led her to hypothesize that aspirin might have significant therapeutic value as part of a breast cancer treatment protocol.[16] Animal studies and statistical analyses of patients' records point in that direction, but she has been unable to raise the funds necessary to prove her case. "A drug that could be patented would get a randomized trial," she laments, "but aspirin, which has amazing properties, goes unexplored because it's 99 cents at CVS."[17]

The American public has been conditioned to believe that cancer treatments are expensive because potential cures are inherently esoteric and elusive, but that is untrue. They are expensive because Big Pharma demands megaprofits to produce them.

Big Pharma and the Third World

The Third World[18] has not historically been a priority for Big Pharma, as impoverished people in poor countries present limited market opportunities.

By late 2014, the Ebola virus had stricken more than thirteen thousand people and killed almost five thousand, mostly in three African countries: Guinea, Liberia, and Sierra Leone. Yet the pharmaceutical industry had shown little interest in Ebola-related research. "A profit-driven industry does not invest in products for markets that cannot pay," lamented Dr. Margaret Chan, director general of the World Health Organization. The WHO, she added, "had long warned of the consequences of greed in drug development and of neglect in public health." So long as Ebola was "confined to impoverished African countries," she explained, "there was no incentive to develop a vaccine."[19] Only when the virus threatened to spread to the industrialized countries did the drug manufacturers begin to consider trying to combat it.

"Intellectual Property Rights" and Monopoly Pricing

Not only do the major drug corporations fail to use their resources to prevent or cure Third World diseases, they have also used them to fight against the efforts of others to do so. When the WHO proposes improvements in international health practices, aimed at providing medicines to people who can't pay for them, Big Pharma mobilizes its lobbyists and financial power to defeat the proposals.

Above all, the drug giants fear changes threatening the international patent-monopoly system that enables them to reap superprofits. Big Pharma's argument in defense of its patents goes like this: "Intellectual property protection is a vital part of health care innovation, providing necessary incentives for investment in research to create new treatments which can help people around the world." That is how Andrew Witty, CEO of GlaxoSmithKline, presents the case.[20] As reasonable as that may sound, the intellectual property rights claimed by companies like

GlaxoSmithKline are often in sharp conflict with a fundamental human right—the right to access to existing medicines of life-and-death necessity for billions of Third World patients.

In 1997, at the height of the worldwide AIDS crisis, South Africa passed a law—the Medicines Act—allowing the government to bypass patent restrictions on antiviral drugs to treat the HIV epidemic in that country. The patented drugs cost from $10,000 to $15,000 per patient per year, which was far beyond the means of the vast majority of South Africa's citizens. The Medicines Act permitted the production or import of generic versions of the drugs at much more affordable prices. In early 2001, forty multinational pharmaceutical firms sued South Africa, demanding that it recognize and uphold their patents. Fortunately, a firestorm of international public health rights protests forced the drug companies to back down. The governments of Germany, France, the Netherlands, and eventually even the United States supported the South Africans, and in April 2001 the lawsuit was dropped.

In India, where some 97 percent of the population lives on about five dollars a day, the pharma giant Bayer was charging $4,500 for a month's dosage of a kidney and liver medicine. In response, in 2012 the Indian government issued a license to another manufacturer to produce a generic alternative with a price tag of $135 a month. The license required the generic producer to pay royalties to Bayer. All of this was legal under laws passed by the Indian Parliament and upheld by its Supreme Court. Big Pharma's response was all-out lobbying and litigation to challenge those laws, but India has not backed down.[21] A major goal of the pharmaceutical industry's challenge has been to overturn the prohibition of "evergreening."

Evergreen Patents

Big Pharma says its patents are justified by the high cost of research and development involved in discovering new medicines. "This just isn't so," declares Joseph Stiglitz, a Nobel laureate in economics:

> For one thing, drug companies spend more on marketing and advertising than on new ideas. Overly restrictive intellectual property rights actually slow new discoveries, by making it more difficult

for scientists to build on the research of others and by choking off the exchange of ideas that is critical to innovation. As it is, most of the important innovations come out of our universities and re-search centers, like the National Institutes of Health, funded by government and foundations.[22]

Nevertheless, the world's governments have generally allowed drug firms to patent the fruits of their research, usually for a period of twenty years. But that seemingly generous concession has not satisfied the man-ufacturers. They want the right to *permanently* charge monopoly prices for their products.

A primary tactic they devised to keep their patents "evergreen" is to make small, insignificant modifications to a medicine's formulation just before its patent is due to expire, and then repatent the "new, improved version" for another twenty years. Governments of underdeveloped coun-tries quickly got wise to this ploy and have vigorously fought against it. In-dia, in particular, has been in the forefront of the resistance. Section 3(d) of India's Patents Act disallows patents for "the mere discovery of a new form of a known substance which does not result in the enhancement of the known efficacy of that substance."

In 2007, Novartis was selling a patented drug in India to treat leuke-mia for $3,000 per dose. Meanwhile, a generic version of the same drug could have been produced for $200 a dose. When the Indian patent of-fice rebuffed Novartis's application to evergreen its patent, the company initiated a legal challenge that eventually reached the Indian Supreme Court, which in 2013 rejected Novartis's final appeal. That landmark judgment "was praised by public-health activists, who said it would protect India's ability to make inexpensive generics for the developing world."[23] A spokesperson for the medical relief group Médecins Sans Frontières says that India, which exports more than half of its $10 billion generic medicine production, is "literally the lifeline of patients in the developing world."[24]

Big Pharma Shifts Gears

The decision by India's Supreme Court was a victory for the public good, but it was just one battle in an ongoing war. The struggle between pa-tients' well-being and monopolistic drug pricing continues. Big Pharma adjusted its strategy from head-on attack to a stance of compromise, but its goal—profit maximization—remains the same.

In 2015, Novartis offered to make a number of its medicines available to Ethiopia, Kenya, and Vietnam at sharply reduced prices. This was undeniably a positive development for people in those countries who desperately needed the drugs, but as one pro-business commentator noted, "Novartis gains too." The company "believes it can still make enough money across its whole product offering and geographic range."[25] The change in strategy was outlined in a 2010 position paper issued by McKinsey and Company, a major business consulting firm to the pharmaceutical industry.[26] It began by noting, correctly, that Big Pharma had long been "paying scant attention to emerging markets and the diseases more prevalent there." (Euphemism alert: In McKinsey's lexicon, poor countries are identified as "emerging markets" and wealthy ones as "developed markets.") However, the paper continued, "emerging markets contributed 30 percent of the pharmaceutical industry's value in 2008, and their pharma markets are forecast to grow by 14 percent a year to 2013."

One tactic for tapping those markets is "tiered" or "differential" pricing, an extreme example of which is a hepatitis-C medication named Sovaldi. A single Sovaldi pill that sells for $1,000 in the United States costs $10 in India. Despite the 99 percent discount, the manufacturer, Gilead, still stood to gain about $1,800 in revenue from each Indian patient who took a full course of Sovaldi.[27]

Unethical Research in the Third World

The McKinsey position paper also advised the drug firms to establish "low-cost research establishments in emerging markets" in order to reduce the expense of drug trials involving large numbers of human subjects. That would also avoid the "stricter regulatory environment, and increasing levels of patent litigation" of the "developed markets."

A GlaxoSmithKline spokesman observed that "trials in Eastern Europe, Asia and Central and South America might cost 10% to 50% less than trials in the USA and Western Europe."[28] That is because safety regulations and their enforcement are much less stringent. This rapidly growing cost-cutting trend exacerbated an already existing bad practice: the widespread use of vulnerable populations as unwitting subjects in clinical drug trials.

John Le Carré's popular novel *The Constant Gardener* was based on a notorious case in which the Pfizer corporation tested a new antibiotic on Nigerian children without the consent of their families or authorization of the Nigerian government. The unethical clinical trials occurred in 1996 and gained public attention in 2000, but did not come fully to light until a suppressed official report was anonymously leaked to the *Washington Post* in 2006. US congressman Tom Lantos declared, "I think it borders on the criminal that the large pharmaceutical companies, both here and in Europe, are using these poor, illiterate and uninformed people as guinea pigs."[29]

The Paradox of Philanthrocapitalism

The Bill and Melinda Gates Foundation has been a major enabler of the burgeoning drug research in the Third World. The Gates Foundation is generally perceived as an agent of positive social change for its philanthropic work, but critics point to it as the prime example of "philanthrocapitalism." Its $50.7 billion endowment makes it the largest private foundation in the world.[30] As such, it is itself a major investor, and Big Pharma firms such as Pfizer, GlaxoSmithKline, Merck, and Johnson & Johnson make up a significant part of its portfolio.

There is, however, an essential difference between philanthrocapitalist foundations and the blatant tax dodges established by right-wing billionaires to fund anti-science propaganda. The Gates Foundation—as one example—has invested $200 million in an effort to provide the world with something most Americans take for granted: a sanitary way to dispose of their urine and feces. More than half the world's population—4.5 billion people, according to UNICEF—does not have access to sanitary toilets, and infections caused by poor sanitation result in the deaths by diarrhea of 480,000 children every year.[31] Despite the free-market fetters attached to the Gateses' "toilet revolution," they have at least made an effort to tackle an appallingly neglected global health problem.

A Gates Initiative to Aid Impoverished Women Worldwide

The Gates Foundation has also sponsored laudable efforts to extend reproductive health care access to women in underdeveloped countries. In 2012, Melinda Gates announced a $4 billion plan "to get 120 million more women access to contraceptives by 2020."[32] The particular contraceptive at issue was Pfizer's Depo-Provera. The initiative was important, because it offered many of the world's poorest women a measure of control over their lives by helping them escape the cycle of constant childbearing and childraising.

That is not the whole story, however. *MarketWatch*, a Dow Jones subsidiary, commented that "$4 billion in new research for women's health care makes Melinda Gates perhaps the biggest player in the future of pharmaceuticals worldwide," and added:

> So you do the numbers: If 120 million new women users chose Depo-Provera, at an estimated average cost between $120–$300 per woman annually, that works out to $15 billion to $36 billion in new sales annually, a nice payoff from leveraging $4 billion in research money.[33]

What this means is that by enormously expanding the market for Depo-Provera, the Gates Foundation enables Pfizer to reap windfall profits. That, in turn, increases the value of the Gates Foundation's stake in Pfizer, further enriching the foundation, and Bill and Melinda Gates's already outsized share of control of the world's economic resources increases accordingly.

There you have the paradox of philanthrocapitalism in a nutshell. Charitable foundations often provide needed medical aid that negligent governments of rich and poor countries alike fail to provide. But like all charity, it does not address root causes. The endemic health crisis of the Third World stems from its poverty. The Gates Foundation, in conjunction with Big Pharma, *perpetuates and exacerbates* global economic inequality.

Conflicts of Interest in Medical Research

In late 2018 the cancer research world was rocked by revelations that Dr. José Baselga, chief medical officer at Memorial Sloan Kettering Cancer Center, had taken millions of dollars from pharmaceutical firms without disclosing the fact. In more than a hundred research articles he had authored since 2013, he failed to mention his extensive corporate connections. In one especially egregious example, in 2017, he put a positive spin on the results of two Roche-sponsored clinical trials that many others considered disappointments, without disclosing his relationship to the company. Since 2014, he has received more than $3 million from Roche.[34]

When Dr. Baselga's transgressions came to light, he resigned from Memorial Sloan Kettering in disgrace. His downfall should not have been all that surprising. As a *New York Times* editorial observed, his fall from grace

> illuminated a longstanding problem of modern medicine: Potentially corrupting payments by drug and medical device makers to influential people at research hospitals are far more common than either side publicly acknowledges.[35]

The corporations bear primary responsibility for corrupting science, but they don't do it alone. Medical ethicist Marcia Angell contends that the main moral onus is not on the businessmen, *even if they initiate and control the research,* but on doctors and scientists like Dr. Baselga who collude with them:

> It is simply no longer possible to believe much of the clinical research that is published, or to rely on the judgment of trusted physicians or authoritative medical guidelines. I take no pleasure in this conclusion, which I reached slowly and reluctantly over my two decades as an editor of the *New England Journal of Medicine.*[36]

"As reprehensible as many industry practices are, I believe the behavior of much of the medical profession is even more culpable," Dr. Angell writes. Although courts of law have from time to time found most major drug companies guilty of fraud (she lists GlaxoSmithKline, Pfizer, Merck, Eli Lilly, Abbott, and TAP Pharmaceuticals as examples), the corporations were merely doing what corporations do—trying to maximize their investors' return. But "physicians, medical schools, and professional organizations have no such excuse, since their only fiduciary responsibility is to patients."[37]

The scope of the corruption is enormous. The total amount Big Pharma pays doctors and medical researchers cannot be directly calculated, but Dr. Angell estimates, by analyzing the annual reports of the nine largest drug firms, that it amounts to tens of billions of dollars a year. And what they have received in return is drug marketing campaigns masquerading as science.[38] In recent years, the conflicts of interest have become so blatant and obvious that the universities and medical schools have been embarrassed and compelled to respond. Their response, however, has been pathetically weak. "They consistently refer to 'potential' conflicts of interest, as though that were different from the real thing, and about disclosing and 'managing' them, not about prohibiting them." What they want is to "eliminate the smell of corruption, while keeping the money."[39]

The Nemeroff Case

Another case study illustrates just how noxious that odor can be. It involves GlaxoSmithKline, the National Institute of Mental Health, and a luminary of American psychiatry, Dr. Charles B. Nemeroff. The depth of their collusion was revealed in the course of a 2008 investigation by the US Senate Committee on Finance.

In 2003, Nemeroff was awarded a $3.95 million grant from the National Institute of Mental Health for a five-year study of five GlaxoSmithKline antidepressant drugs. Of the grant money, $1.35 million went to Emory University, where Professor Nemeroff was chair of the psychiatry department. This study obviously should have been rigorously shielded against any influence on the part of GlaxoSmithKline, but that was not the case. Nemeroff was on Glaxo's payroll and everyone knew it. He was not required to distance himself from the company but simply to disclose to Emory University how much Glaxo was paying him, and if it exceeded $10,000 a year, Emory was supposed to report that to the National Institutes of Health.

Nemeroff agreed to those terms but failed to abide by them. He didn't report hundreds of thousands of dollars he had received from Glaxo for giving talks promoting Paxil, an antidepressant, among other Glaxo products. When Emory discovered that he wasn't upholding the agreement and asked for an explanation, he denied all wrongdoing but said that from that time forward he would not accept more than $10,000 a year from Glaxo.

He lied again. In 2004, Glaxo paid him $171,031, and he reported to Emory that he'd received . . . (wait for it) . . . $9,999. Emory should not have been very surprised by this. Earlier, Nemeroff had written to its financial administrators to remind them, with breathtaking arrogance, of how much Big Pharma money he was bringing to the university:

> Surely you remember that Smith-Kline Beecham Pharmaceuticals donated an endowed chair to the department and there is some reasonable likelihood that Janssen Pharmaceuticals will do so as well. In addition, Wyeth-Ayerst Pharmaceuticals has funded a Research Career Development Award program in the department, and I have asked both AstraZeneca Pharmaceuticals and Bristol Meyers [*sic*] Squibb to do the same. Part of the rationale for their funding our faculty in such a manner would be my service on these boards.[40]

Charles "Paxil" Nemeroff became the poster boy for conflict of interest in medical research. Glaxo was certainly not his only client; at one point in his storied career he had consulting arrangements with *twenty-one* drug firms simultaneously.[41]

After the 2008 congressional investigation revealed that Nemeroff had taken more than $2.8 million from drug firms between 2000 and 2007 and neglected to report at least $1.2 million of it to Emory, he resigned his department chairmanship but remained on the faculty. He was prohibited from taking part in sponsored research grants, but only for two years. The brevity of that ban suggests the extent to which Emory, no less than Nemeroff, was addicted to Big Pharma money. Nonetheless, by 2010 Nemeroff had moved on to greener pastures. The University of Miami overlooked his ethically challenged past and invited him to chair its psychiatry department.

As a postscript to this affair, in 2012 GlaxoSmithKline pleaded guilty to criminal charges under the False Claims Act and agreed to pay the US government a $3 billion fine—the largest settlement any pharmaceutical company had ever made. Nonetheless, even multibillion-dollar fines are insufficient to effectively deter Big Pharma's fraudulent practices. In this case, Glaxo had been fined for activities relating to three of its products: Paxil, Avandia, and Wellbutrin.

> $3 billion represents only a portion of what Glaxo made on the drugs. Avandia, for example, racked up $10.4 billion in sales, Paxil

brought in $11.6 billion, and Wellbutrin sales were $5.9 billion during the years covered by the settlement.[42]

Glaxo could well afford to write off the $3 billion fine as just another cost of doing business.

The Nemeroff case was by no means an anomaly. If Nemeroff is the grand prize–winner of the Conflict-of-Interest Sweepstakes, the first runner-up is Harvard child psychologist Joseph "Your Child Is Probably Bipolar" Biederman. A fortyfold increase in pediatric bipolar disorder diagnoses has been attributed in part to Dr. Biederman's cheerleading for Johnson & Johnson's antipsychotic drug Risperdal.[43] Dr. Biederman was found by the aforementioned 2008 congressional investigation to have "consulted" for the drug industry to the tune of at least $1.6 million while only reporting $200,000 of it to Harvard.[44] Nonetheless, as of this writing Dr. Biederman is still at Harvard and still in charge of pediatric psychopharmacology research at Massachusetts General Hospital.

A Turning Point: The Bayh-Dole Act

Drs. Nemeroff and Biederman are high-profile personifications of how academic research institutions have come to disregard conflicts of interest. It has not always been so extreme. A pivotal change occurred in 1980, when a new law, the Bayh-Dole Act, allowed universities to patent discoveries funded by taxpayer dollars. The legislation's sponsors said it would provide incentives for innovation in academic research, but it proved instead to be a slippery slope to the debasement of the university research system. It provoked, as one critic commented, "an intellectual property gold rush."[45] By removing public-license restrictions, Bayh-Dole opened the floodgates to corporate investors seeking monopoly ownership of innovative technology.

The act originally permitted only universities and small businesses to own patents derived from federally funded research, but as surely as night follows day, major corporations were bound to eventually gain the same privilege. In 1987—only seven years later—they did. The upshot has been the blurring of boundary lines separating government, university, and industry research. Pharma giants like Pfizer or Glaxo leverage their research investments by combining them with federal funding of medical school laboratories. For their contribution, they gain exclusive licensing of the fruits of the research. In other words, *public dollars now routinely pay universities to produce knowledge that becomes the private intellectual property of corporations.*

Conflicted Patient Advocacy Groups

Patient advocacy groups provide another fertile field for pharma corruption. Organizations such as the Leukemia & Lymphoma Society, the American Diabetes Association, Food Allergy Research & Education, and the National Multiple Sclerosis Society all receive a significant portion of their funding from pharmaceutical firms. An umbrella group representing such organizations, the National Health Council, gets 62 percent of its $3.5 million budget from drug companies. Dr. Michael Carome of Public Citizen understated the obvious: "It is a conflict of interest."[46]

The patient advocacy groups do nothing illegal by taking money from Big Pharma, but their pervasive conflicts of interest parallel those that have undermined the integrity of medical science. Even the World Health Organization is not immune. Thirty percent of WHO's 2011–2012 budget, $4.9 billion, was from private donors or grants from governments over which Big Pharma exerts considerable influence. "These contributions are earmarked for specific purposes, allowing the donors to directly influence WHO's work," says Thomas Gebauer of the medical relief organization Medico International.[47]

The Strategy and Tactics of Corrupting Medical Science

Big Pharma's marketing success depends upon creating the illusion that its products' therapeutic value rests on a solid foundation of scientific authority. Subordinate to that general strategic orientation are a number of standard tactical maneuvers. The primary tactic is the grooming of what the industry calls "KOLs," or key opinion leaders. A less euphemistic description of the practice would be "hiring scientists as pitchmen." Glaxo's utilization of Charles Nemeroff is but one example of that tactic in action.

A few of the other standard operating malpractices are known by the terms "seeding trials," "medical ghostwriting," "predatory journals," and "disease mongering." Pondering these various stratagems leads to the reflection that if only the ingenuity devoted to public relations and marketing had instead been put into scientific research, the human race's health problems would certainly be far more manageable by now.

Seeding Trials

A "seeding trial" is designed to look like a clinical drug trial but is in fact a marketing ploy conducted by a pharmaceutical company. Its purpose is not to study a new drug's efficacy or safety but to introduce a new medicine to doctors, right as it enters the market, so they will be more likely to prescribe it to their patients. In short, it is a bogus study with no scientific merit.

It is an ingeniously deceptive ruse. Doctors who participate are led to believe that their patients are the trial subjects when it is actually the doctors themselves who are the subjects of a behavioral modification experiment. The practice came to light thanks to litigation that publicized some internal corporate documents involving two brand-name pain medications—Merck's Vioxx and Pfizer's Neurontin. Merck's study, in which three subjects died and five more had heart attacks, was entirely designed and conducted by the company's marketing department. Pfizer's study

> was notable for how poorly it was conducted. The investigators were inexperienced and untrained, and the design of the study was so flawed it generated few if any useful conclusions. Even more alarming, 11 patients in the study died and 73 more experienced "serious adverse events."[48]

Bioethicist Carl Elliott asks, "How can studies that endanger human subjects attract so little scrutiny?"[49] In an earlier era, when most clinical research was conducted in university laboratories, there was a substantial regulatory system designed for the protection of human subjects. In the new era of "private sector" research, it is no longer adequate.

The limited protection that does exist for human subjects in medical research is provided by a system of small committees called Institutional Review Boards (IRBs). The IRBs derive their mandate from a number of federal agencies. They don't monitor the research directly; they merely review the trial protocols before the research is performed. The most glaring flaw in the IRB system is that many of them are privately owned businesses that are paid by the pharma firms they are supposedly monitoring.[50] If an IRB were actually to challenge a study, the researchers could go shopping for a more compliant one. IRBs therefore have a financial incentive to rubber-stamp Big Pharma's seeding trials.

Medical Ghostwriting

Just because an eminent scientist's name is listed as an author of an article in an authoritative science journal, it cannot be assumed that he or she had anything to do with actually writing it, or with the research it reports. Ghostwriting has become a substantial problem in scientific and medical publishing. The scientists who lend their names to articles they haven't written often haven't even seen the data the articles are based on. "Some of us believe that the present system is approaching a high-class form of professional prostitution," laments E. Fuller Torrey, a research director of the Stanley Medical Research Institute and an esteemed expert on schizophrenia.[51]

Even worse than the companies that merely provide ghostwriting services are those firms that do both—not only writing the article but also performing the clinical trials themselves. One prominent example is Scirex, "a little-known research firm owned partly by Omnicom, one of the world's biggest advertising companies." Omnicom and two other advertising giants, Interpublic and WPP, "have spent tens of millions of dollars to buy or invest in companies like Scirex that perform clinical trials of experimental drugs."[52]

From the beginning of the 1990s to the end of that decade, the proportion of pharma industry funding that went to university laboratories declined from three-quarters to one-third, with the rest going to in-house researchers or to ad-agency outfits like Scirex. What this means, a former editor of the *New England Journal of Medicine* says, is that "you cannot separate their advertising and marketing from the science anymore."[53]

The most prestigious medical journals make an effort to bar ghostwriting and at least try to acknowledge their authors' financial conflicts of interests, but the journals themselves are far from conflict free. A 2007 investigation revealed that the two leading American medical journals, the *New England Journal of Medicine* and *JAMA, the Journal of the American Medical Association,* take in a substantial portion of their budgets, $18 million and $27 million a year, respectively, in pharmaceutical advertising revenue.[54]

The conclusion reached by Richard Horton, long-time editor of the *Lancet*, seems beyond dispute: "Journals have devolved into information laundering operations for the pharmaceutical industry."[55] But when he issued that negative appraisal of medical journals in 2004, he was writing about the best of them—the ones that had made the greatest efforts to maintain their integrity.

A Brief Digression: The Lancet *and the Anti-vax Movement*

The *Lancet* is among the most trustworthy of the medical journals, but it nonetheless was the vehicle that greatly amplified one of the more troubling episodes in recent medical science history, the anti-vaccination mania. In February 1998, the *Lancet* published a paper by a Dr. Andrew Wakefield presenting evidence that widely used vaccines could be the cause of autism in children.[56]

MMR vaccines had been in use for decades as an effective defense against childhood mumps, measles, and rubella. Wakefield's article triggered a widespread protest movement of frightened parents in opposition not only to MMRs but to the use of vaccines in general. Wakefield's claims were thoroughly debunked by responsible medical researchers. The evidence is unambiguous and overwhelming: there is no link between vaccines and autism.[57] Wakefield himself was exposed as having fabricated his data and having financial conflicts of interest to boot.[58] In 2010, the *Lancet* officially retracted the study. But it was too late; the anti-vax movement had taken on a life of its own.

In Britain, where the *Lancet* is published, the repudiation of Wakefield's article dealt a setback both to Wakefield himself and to the British anti-vax movement. But Wakefield was not to be deterred. He moved his field of operations to the United States and gained celebrity status by exploiting the anti-science and conspiracy-theory passions of Donald Trump's supporters.[59] Sadly, the American anti-vax movement has spread "vaccine hesitancy" to the point where the number of unvaccinated schoolchildren has begun to seriously undermine public health.

The World Health Organization has declared vaccine hesitancy one of the ten biggest threats to global health. When the proportion of vaccinated members of populations dips below 95 percent, the "herd immunity" necessary to prevent disease outbreaks is lost. In the United States in 2017, only 91.5 percent of children from nineteen to thirty-five months of age received the MMR vaccine.[60] In April 2019, the Centers for Disease Control reported 695 cases of measles in twenty-two states, "the greatest number of cases reported in the United States since measles was eliminated from this country in 2000."[61] One of the unfortunate aspects of the anti-vax agitation has been a demagogic effort to inflame Trump's followers by playing on their legitimate antipathy toward Big Pharma. The pharmaceutical industry thus gains undeserved credit for being on the side of science and truth in this contrived controversy.

Predatory Journals

Meanwhile, the race to the bottom accelerated with the advent of the "fee-based open-access" journal. Open-access scientific journals made their debut in the late 1990s and increased sharply in number in the first decade of the new century. There are about ten thousand of them now, and many are devoted to the medical sciences. They often bear names similar to those of established publications (*Les cahiers des sciences naturelles* instead of *Cahiers des sciences naturelles*, for example).

The original idea behind the open-access movement was to make scientific information more widely available by not charging readers a fee to read it. Instead, the authors of scientific studies would pay the costs of having their own papers published. At first, a number of legitimate open-access journals attempted to maintain high scientific standards requiring rigorous peer review. But then less scrupulous publishers discovered they could charge the same high fees without the expense of ensuring the quality of their product. They would simply publish any research paper submitted to them with no oversight at all, as long as the author paid the fee. These bogus journals were invariably online publications, so their production costs were negligible.

Before long the internet was awash with impressive-looking but worthless studies masquerading as science. In 2014 an estimated 420,000 papers were published in predatory journals.[62] Gresham's Law applies to science as well as to money: "Bad science drives out good." As distinguishing legitimate research from junk science becomes increasingly time-consuming, the body of scientific literature is ever more contaminated. The consequences of a flood of untrustworthy results are particularly ominous for medical science (and by extension, for human health).

The success of the predatory journals was made possible by the collusion of their prey. As investigative science journalist Gina Kolata discovered, "Many—probably most—academics who publish in these journals know exactly what they are doing. They are padding their résumés, taking advantage of the fact that colleges may not know if a journal is legitimate or not."[63] It amounts, she declares, to "academic fraud" that "chips away at scientific credibility, and muddies important research."[64]

The phony journals have given rise to a wider assortment of fraudulent science institutions. Authors who publish papers are then invited—for an additional fee—to present them at academic conferences with impressive titles. That looks good on their résumés, and they don't even have to attend the sham conferences.

Disease Mongering

In the beginning was the disease. A medicine to treat it would come later. That was the natural, logical order of things, but Big Pharma has turned it on its head. As Dr. Angell observes, "Instead of promoting drugs to treat diseases, they have begun to promote diseases to fit their drugs."[65]

This ass-backwards modus operandi is not new. It may have had antecedents, but it began in earnest in the 1990s, in the aftermath of the Prozac revolution. The phrase now commonly used to identify it, "disease mongering," first appeared in the title of a 1992 book on the subject.[66]

Prozac (Eli Lilly's brand name for fluoxetine) was the first of a class of antidepressants called SSRIs to win FDA approval.[67] It became the trendiest of the mood-modifying drugs of the 1990s. Other members of the group included Paxil, Zoloft, and Celexa, all of which joined the "blockbuster club" by reaching the billion-dollar level in annual sales. Between 1985 and 2007 the total annual sales of antidepressants in the United States increased fiftyfold, from $240 million to $12 billion.

Big Pharma, however, discovered an ingenious way to expand the already vast market for its SSRIs. Prozac was approved as an antidepressant, but an ambiguity in FDA regulations allowed it, with minimal additional testing, to be marketed for other purposes. The industry exploited that loophole to the utmost by searching for (and, when necessary, inventing) "other purposes" and publicizing them to the skies. Eli Lilly invented another purpose for Prozac by changing its brand name to Sarafem (thereby extending its patent by seven years) and promoting it as a treatment for a condition called pre-menstrual dysphoric disorder, or PMDD. The first Sarafem commercial explicitly differentiated the new, exotic PMDD from boring, familiar PMS.[68]

The industry's playbook "is almost mechanized by now," says Dr. Loren Mosher, a former National Institute of Mental Health official. It begins with a public relations campaign to create "awareness" of the disease to be promoted, focusing on "a mild psychiatric condition with a large pool of potential sufferers." Then, to satisfy the minimum FDA requirement, a small hypothesis-driven study will establish that the drug can be useful in treating the disease. If the first study fails to produce the desired result, the researchers will try, try again until they get one that confirms what they set out to find.[69]

The PR campaign will then shift focus from the disease to the drug. It will feature authoritative MDs who are recruited to sing the medicine's praises, and cite impressive statistics generated by the firm's own tailored research. To cap it off, a patient advocacy group—its funding sources carefully hidden—will be created as a front group to promote the drug. *Et voilà!* A new disease is born, accompanied by just the right pill to cure it.

Remember Paxil, the drug promoted by Dr. Nemeroff? Paxil was a GlaxoSmithKline antidepressant, but in 1999 the FDA gave Glaxo approval to market Paxil as a treatment for social anxiety disorder. Paxil's product director boasted, "Every marketer's dream is to find an unidentified or unknown market and develop it. That's what we were able to do with social anxiety disorder."[70]

This is a peculiarly American problem. It is the consequence of a specific enabler of Big Pharma's PR campaigns that exists only in the United States and New Zealand: direct-to-consumer drug advertising. In 1997 the FDA gave the pharma firms the right to run television commercials for prescription medicines. It wasn't long before the brand names of blockbuster drugs like Prozac, Viagra, and Claritin had become household names.[71]

What all the new syndromes, disorders, and dysfunctions have in common is that they medicalize and pathologize aspects of everyday life (menstruation,[72] menopause,[73] teenage moodiness[74]), or transform mild conditions (twitchy limbs[75]) into serious diseases. They proliferate so fast that it is difficult to keep up with them, but a short list includes social anxiety disorder, general anxiety disorder, pre-menstrual dysphoric disorder, restless legs syndrome, irritable bowel syndrome, and seasonal affective disorder. Many of the conditions have a sexual dimension: erectile dysfunction, female sexual dysfunction, testosterone deficiency, and hypoactive sexual desire disorder. It is not my intent to suggest that any or all of these conditions are not real or are not serious. The point is that our consciousness of them is almost entirely a product of the pharma industry's public relations machine.

Post-9/11 Syndrome?

The vast majority of the mongered diseases were created or inflated to sell psychoactive drugs, because marketers know that subconscious fears and anxieties are especially easy to manipulate. A particularly vile example was Pfizer's promotion of Zoloft to treat post-traumatic stress disorder in the wake of the September 11, 2001, World Trade Center attack.

Pfizer's PR agency, Chandler Chicco, had already created a patient advocacy group, the PTSD Alliance, to publicize the new malady. On September 26, the front group proclaimed that those who had "witnessed a violent act" or experienced "natural disasters or other unexpected, catastrophic, or psychologically distressing events such as the September 11 terrorist attacks" were at risk for PTSD.[76] That would presumably include the hundreds of millions who "witnessed" the event on television.

Not to be outdone, GlaxoSmithKline upped the ante with widely broadcast post-9/11 commercials for Paxil, featuring sad faces saying things like, "I'm always thinking something terrible is going to happen" and "It's like a tape in my mind. It just goes over and over and over."[77] Pfizer and Glaxo thus used PTSD to bolster the vastly exaggerated fears of terrorism that have been exploited to justify endless wars abroad and erode civil liberties at home.

Oversight and Regulation: The FDA

It should be abundantly clear by now that pharmacology is not the only medical or biochemical science corrupted by Big Pharma. Psychiatry,

oncology, cardiology, and all the rest—none have been left unsullied by·
the pervasive corporate conflicts of interest.

Could anything be done to restore the integrity of the medical sciences? Yes, if oversight and control in the public interest could be exercised by the governmental agencies mandated to do so: the Food and Drug Administration, the Environmental Protection Agency, and the Federal Trade Commission's Bureau of Consumer Protection, for starters. But how can an agency like the Food and Drug Administration be expected to cope with the corrupting influence of Big Pharma when it itself is beholden to the industry? As medical ethicist Sheldon Krimsky declared, the FDA long ago devolved into "an agency where conflicts of interest have become normalized in the process of drug evaluation."[78] In 2000 an investigation disclosed that "more than half of the experts hired to advise the government on the safety and effectiveness of medicine have financial relationships with the pharmaceutical companies that will be helped or hurt by their decisions."[79]

In November 2004, David Graham, an FDA safety officer for twenty years, testified before the US Senate that his supervisors had tried to "silence him and pressure him to limit his criticism of the safety of some drugs." Graham's allegations were corroborated the following month by a report of the Health and Human Services Department's inspector general, which concluded that "the work environment at the FDA's Center for Drug Evaluation and Research either allowed little dissent or stifled scientific dissent entirely." Of 360 FDA scientists surveyed, 63 reported having been "pressured to approve or recommend approval for a (new drug application) despite reservations about the safety, efficacy or quality of the drug."[80]

But the FDA's incestuous relationship with Big Pharma goes much deeper than mere conflicts of interest and bureaucratic muzzling of honest scientists. The public expects the regulators to have the upper hand in that relationship, but the muscle is all on the corporate side: the FDA's entire budget for drug regulation amounts only to about 1 percent of what the drug industry spends annually for marketing.[81]

Where does the money for that FDA budget come from? It "relies extensively on user fees," says Dr. Steven Nissen, chair of the Department of Cardiovascular Medicine at the Cleveland Clinic Foundation. The drug firms applying for marketing approval pay the user fees to the FDA, which means that "the FDA is financially indebted to the companies it must regulate."[82] Another critic states the case more harshly: user fees are "little more than a form of legalized bribery that all but guarantees a favorable response every time a pharmaceutical company submits a new product for review."[83] These fees are not small change; they amount to billions of dollars. In 2016 the application fee for any prescription drug application, including clinical data, was a minimum of $2.3 million. Between 1992 and 2016, user fees paid by pharmaceutical companies to the FDA totaled $7.67 billion.[84]

In addition to approving new drugs, the FDA also grants the applicants exclusive marketing rights, which together with the patent system are the bulwarks of the industry's monopoly pricing system. When the pharma firms are resisting all regulation, its lobbyists will hypocritically pontificate about their devotion to "free enterprise." Big Pharma's super-profits, however, depend upon the *unfree* market enabled by the exclusive marketing rights the FDA grants them.

A Few Modest Proposals

Again, can anything be done to restore the integrity and credibility of American medical science? Yes. The solution, in two words, is *public control*. Or, in four words that are anathema to Big Pharma and its political allies, *federal regulation with teeth*.

First, all medical research should be publicly funded. Second, it should be vigorously regulated to eliminate all financial conflicts of interest. That would require excluding any and all interference by private corporate interests. Privatized science is an oxymoron. Scientists on corporate payrolls are pressured to selectively report their findings by suppressing data that conflict with their patrons' marketing plans. Eliminating these practices does not mean limiting the amount corporations can pay scientists studying their products to $10,000 a year. It means limiting the amount to *zero*.

Regulatory agencies like the FDA must be adequately funded from tax revenues rather than forced into dependence on "user fees" paid by those they are regulating. Their administrators and staffs should have no connections to the drug industry. Private corporations must not be allowed to patent discoveries made with public funds. Direct-to-consumer advertising of medicines must be banned, as it is virtually everywhere else in the world.

Those who call themselves political realists will object that these proposals are impossible to achieve in the current American political context. They are not mistaken. The choice, then, is between science and the current American political context.

CHAPTER 6

Spitting in the Well We Drink From

IN FEBRUARY 1979 PRESIDENT CARTER made a notable faux pas. At a public gathering in his honor in Mexico City, he began a toast to the Mexican president with a joke about "Montezuma's revenge." It got a nervous laugh, but his audience was astounded that the affable American president had no idea how deeply insulting that phrase is to the Mexican people.

"Montezuma's revenge" is an allusion to the diarrhea that tourists contract in less affluent countries by drinking unfiltered tap water. British tourists in India call it "Gandhi's revenge" and "Delhi belly." Travelers to all Third World countries are routinely warned: "Don't drink the water!" All of this reflects an imperialistic insensitivity to the fact that for most of the world's population, safe drinking water is still an unaffordable luxury.

So imagine how shocked the citizens of Flint, Michigan, must have been when in 2015 its municipal government announced that the city's water was unfit for human consumption and had been poisoning their children for months, possibly years. The revelation had been forced by the research of Dr. Mona Hanna-Attisha, a public health advocate who demonstrated that Flint's children had been exposed to dangerous levels of lead in their drinking water. "As a pediatrician," Dr. Hanna-Attisha says, "I know lead is the worst kind of poison. Permanent. Life-altering. A neurotoxin, lead can have serious consequences on the developing brain."[1]

The crisis was precipitated by years of draconian budget cutbacks that prompted city administrators to spend less on drinking water. To that end, they changed the water source from Lake Huron to the Flint

River, despite that river's history of "poor water quality due to unregulated discharges by industries and municipalities."[2]

Although the river was polluted by organic and inorganic toxins, including fecal coliform bacteria from sewer seepage, public officials "manipulated test results, downplayed the problem, and lied to the residents."[3] After the declaration that Flint's water was undrinkable, the same officials shamefully resisted returning to the Lake Huron supply, arguing that the city simply couldn't afford it. By the end of the year, however, they had been forced to make the reconnection, but it was too late. The crisis continued.

More than one state of emergency was declared, and in January 2016 the National Guard was called out to distribute safe drinking water to the people of Flint. In June 2017, the deadliness of the affair was underscored when the head of Michigan's health department and four other officials were charged with involuntary manslaughter.[4]

Scientific negligence played a significant part in this debacle, and Flint's experience was not without precedent. Marc Edwards, described by the *Washington Post* as "the heroic professor who helped uncover the Flint lead water crisis," had also exposed a similar problem in Washington, DC, a decade earlier. Edwards, a professor of environmental engineering at Virginia Tech, proved

> that corrosion in the nation's capital's pipes had caused lead to seep into the water supply and pass through kitchen faucets and shower heads. After helping to expose that water crisis in 2004, he spent six years challenging the Centers for Disease Control and Prevention [CDC] to admit they weren't being honest about the extent of the damage the lead had on children.[5]

His seemingly quixotic crusade to hold the federal agencies accountable earned him six years of harassment, threats, and ridicule, but finally, in 2010,

> he was vindicated when it was proven that the CDC had lied to the public in a misleading report, which falsely claimed lead levels in the water had not posed a health risk to D.C. residents.[6]

The water crises in Washington, DC, and Flint point to some troubling trends. "We are in an era where the pie is getting smaller," Edwards explained in an interview with *American Scientist*. "That is going to create unprecedented pressure on all aspects of science and engineering."[7] A shrinking federal R&D budget will push American science ever more into the arms of the private sector. Meanwhile, the degradation of privatized science by regulatory agencies in collusion with politicians will lead to environmental deterioration and a fundamental decline in the quality of American life.

The problem has spread far beyond Flint. Pittsburgh and Chicago are likewise battling higher levels of lead in their drinking water, and in August 2019, Newark was plunged into "one of the largest environmental crises in an American city in years."[8] This, Dr. Hanna-Atisha declares, is "one of the legacies of the profit-driven and largely unaccountable lead industry that thwarted science, fought regulations and forced its use in our gasoline, paint and plumbing."[9]

Flint and Newark are local manifestations of a global crisis. The proverbial caveat "Do not spit in the well you drink from" provides a metaphorical warning against what we, as a global society, are doing to ourselves. But it is not strong enough; "poisoning the well" would more accurately reflect our current situation.

Tens of thousands of people—most of them children in poor countries—will die from environmental poisoning in the time it will take you to read this book. Do the math: Cornell University ecologist David Pimentel estimates that 62 million deaths a year can be attributed to water, air, and soil pollution.[10] That's 170,000 deaths a day, worldwide.

Rachel Carson and the Origins of Environmentalism

In the late 1950s, Barry Commoner and other politically conscious scientists founded the St. Louis Committee for Nuclear Information to warn of the dangers of radioactive fallout from atom bomb tests in Nevada. Commoner and his allies spearheaded what they called a science information movement to promote the idea that a scientist's primary obligation is not to governmental policymakers but to the people. It is the moral duty of scientists, they maintained, to fully and directly inform the general public about the scientific aspects of social issues.[11]

The most significant dissident voice to take up that call was Rachel Carson's. The publication of her book *Silent Spring* in 1962 gave rise to the environmentalist movement and sparked a debate that forced a thoroughgoing reconsideration of the place of science in human affairs.

She declared that the massive modern use of pesticides and other synthetic chemicals to produce food posed a serious threat to humankind, and ultimately to the continued existence of all life on earth. By linking an impending environmental crisis to the blind drive for profits by agribusiness and the chemical industry, *Silent Spring* initiated a fundamental challenge to corporate America. The industries she targeted, in collaboration with the US Department of Agriculture, launched a ferocious assault against her and her book. They "set their scientists to analyzing Miss Carson's work, line by line. Other companies are preparing briefs defending the use of their products. Meetings have been held in Washington and New York. Statements are being drafted and counterattacks plotted."[12] Scientists on the payroll of the chemical corporations believed they could dismiss Carson as a mere popularizer who wrote for the public rather than for an audience of elite scientists. But in spite of their attempts to marginalize her, she ignited a momentous social movement. Furthermore, by redirecting interest toward ecology and away from traditional mechanistic and reductionist approaches, *Silent Spring* had a major impact on the way biological knowledge would henceforth be pursued.

Carson's crusade against DDT created a general awareness of the dangers of pesticides. How have the efforts to restrict them fared since? Alas, as important as consciousness-raising is, it has not translated into effective action or lasting solutions. Fifty-four years after *Silent Spring* came off the presses, environmental lawyer and activist Carol Dansereau, in a book entitled *What It Will Take*, assessed the current state of the movement Carson inspired. She compared her own twenty-eight years of experience in defending the public from pesticides and other toxins to playing Whack-a-Mole: "No matter how many times we whack problems down, others pop up to take their place." Battling the "corporatocracy" one toxin at a time, she concluded, is not an effective strategy for preserving the environment. It is the corporate dominance of all aspects of our lives that has to be challenged.[13]

From Nixon to Trump

The rise of environmental consciousness did, however, have a political impact, as evidenced by Richard Nixon's State of the Union message to Congress on January 22, 1970. "Restoring nature to its natural state," Nixon declared, "has become a common cause of all the people of this country," adding for emphasis: "Clean air, clean water, open spaces—these should once again be the birthright of every American."[14] This was not empty rhetoric; he backed it with action and federal funding. "We still think of air as free," Nixon said. "But clean air is not free, and neither is clean water. The price tag on pollution control is high." He proposed a $10 billion clean waters program, created the Environmental Protection Agency in December 1970, and oversaw the enactment of the Clean Air Act of 1970 and Clean Water Act of 1972.

Today, in the era of "the Republican war on science," it seems surreal that a Republican president—especially one who was forced from office in disgrace—could have been responsive to public demands for environmental action, but that was indeed what happened. Eventually, however, the tide would turn and flow in the opposite direction.

When the Trump administration assumed executive power in early 2017, governmental support for numerous anti-science agendas suddenly became significantly more explicit. Among Trump's first official actions was to propose a federal budget that drastically cut support for certain kinds of scientific research; funding for the National Institutes of Health and the Environmental Protection Agency were to be decreased by 18 percent and 31 percent, respectively.[15] The EPA budget cut would require laying off 3,200 of its staff members. Other research fields—those with military applications—fared better in the proposed budget.

Trump and his cronies exacerbated a preexisting condition that had for several decades been nurtured by both major parties' subservience to corporate agendas. The assaults on environmental science have been many and multifarious, but one line of attack stands out as most prominent in the current public discourse.

Denying the Reality of Anthropogenic Global Warming

In 1988 testimony by James E. Hansen, NASA's leading climate scientist, before the US Senate Committee on Energy and Natural Resources "ignited public discussion of global warming and moved the controversy from a largely scientific discussion to a full blown science policy debate."[16] In response, Big Oil and Big Coal tried to follow the playbook established by Big Tobacco, but they have been far less successful than their cigarette-, food-, and pharma-industry counterparts in recruiting credentialed scientists to hoist their banner.

Global warming is not controversial among climate scientists. For at least several decades, the science has been unequivocal: the average temperature of the earth is increasing at an alarming rate. But although there is no scientific debate over the reality of climate change, it has certainly become a contentious political issue. Without the climate scientists' collaboration, the energy corporations have been forced to rely on public relations flimflam and buying political influence. Unfortunately, that has been enough to achieve their main goal: to muddy the national conversation on global warming just enough to give their political allies a fig leaf of an excuse to block effective measures of confronting the problem.

Climate-change denial is promulgated by right-wing politicians, think tanks, and media, but it does not begin with them. They exist only thanks to the generous financial support they receive from Big Oil and Big Coal. These corporations' products provide energy to warm homes, move vehicles, and power heavy industry by burning compounds containing the element carbon. A by-product of that process is carbon dioxide—CO_2—a colorless, odorless gas that enters the earth's atmosphere and remains there. A certain proportion of CO_2 in the atmosphere is to be expected because you, I, and other animals breathe it out every time we exhale. Then the earth's plant life breaks it down and puts the oxygen back in the air for us to inhale again and create more CO_2. That balanced cycle evolved millions of years ago.

But in the past hundred years or so the cycle has begun to become unbalanced, with the proportion of CO_2 in the atmosphere slowly rising. It is now at the highest atmospheric concentration in three million years—since long before humans existed.[17] That matters because CO_2 is a "greenhouse gas," so called because it acts in the atmosphere like the glass in a greenhouse, allowing heat from the sun's rays in but preventing it from going back out. That is the ABC of global warming: the greenhouse effect has brought about a steady increase in the average temperature of our home planet.

The source of the excess CO_2 is no mystery. It is the result of the great increase in the burning of carbon-based fuels—"fossil fuels"—since the rise of the Industrial Revolution in the late eighteenth century. The primary culprits are the coal-fired steam engine and its successors, which turn the wheels of industrial production, and the gasoline-powered internal combustion engine, which turns the wheels of automobiles. Global warming is thus an *anthropogenic* phenomenon—a product of human activity. It stands to reason, then, that if climate change has human causes, humans should also be able to stop the process and reverse it.

The initial danger signs of global warming are evident in shrinking polar ice caps and glaciers. Increasingly erratic weather patterns have produced droughts in some large areas and catastrophic flooding in others. They have triggered record numbers of heat waves, blizzards, wildfires, mudslides, hurricanes, tornados, cyclones, and tsunamis all over the world.[18] A weather disaster that made an especially profound impact on the American psyche was the 2005 hurricane Katrina, which laid waste to a major American city, New Orleans.[19] Hurricane Maria's ruination of Puerto Rico in 2017 and the devastating wildfires that in 2019 and 2020 destroyed vast areas of California, of the Amazon rainforest, and of Australia added more cruel examples to the list of "natural disasters" in which we now recognize an anthropogenic dimension.

The mechanisms driving extreme weather events are highly complex, and many details of the phenomena remain to be studied, but some fundamental processes underlying the increasing frequency and intensity of floods and hurricanes are already well understood. An estimated 93 percent of the heat trapped by the greenhouse effect goes into the oceans.[20] Warmer oceans produce increased evaporation that puts more water vapor in the air, which feeds heavier rainfalls. And warmer air from the oceans colliding with cooler air from elsewhere converts thermal energy into mechanical energy in the form of hundred-mile-an-hour wind-and-rain storms.

As for the long-term consequences, the prognosis is ominous. "Global warming presents the gravest threat to life on Earth in all of human history," declares the Center for Biological Diversity.[21] Scientists are understandably reticent about issuing doomsday predictions, but that ultimate danger is what makes global warming an issue that only the foolishly shortsighted dare to ignore. Measured on the scale of human time perception, the process of climate change is slow and uneven, which gives demagogues who deny its existence a superficial talking point. But climate scientists have demonstrated beyond all doubt that the inexorable *trend* in the temperature of the earth's surface, seas, and air is upward, ever upward.[22]

Government Censorship of Climate Science

The escalating political war against environmental science is reflected in more than a decade of governmental efforts to censor and discredit the climate-change warnings of James Hansen and other scientists. In December 2005, Hansen gave a public lecture on global warming in which he released data revealing that 2005 had been the warmest year in at least a century.

The oil and coal industries were offended, and the Bush-Cheney administration swiftly made its displeasure known. NASA officials warned Hansen of "dire consequences" if he persisted in issuing such statements. He was told that his future lectures and writings, including postings on agency websites, would have to be vetted by NASA "public affairs" personnel (in other words, the PR department), and that he would have to seek approval for any interviews he gave to journalists.

Hansen's treatment was not an isolated case. In a survey of 1,600 scientists at seven government agencies, nearly half reported having been warned against using terms like "global warming" in their reports during George W. Bush's eight years in office. Unlike Hansen, however, most of the other scientists were not in a position to resist the pressures of their administrators.

Political pressures on scientists continued during the Obama administration, but more subtly. In 2009 "a strategic decision by the White House to downplay climate change—avoiding the very word" led Obama to demand of his "allies in the environmental movement" that they engage in self-censorship and keep mum on the issue.[23] To their discredit, many of the mainstream environmentalist organizations agreed to do so, handing the right-wing climate-denialists a large advantage in the public discourse.

Hansen's opposition to governmental climate policies did not end with the replacement of Bush by Obama. He heatedly attacked Obama's "cap and trade" proposals—an attempt to control industrial carbon emissions by the use of market incentives—as a woefully impotent strategy. Such schemes actually create a license to pollute that can be bought and sold. Hansen also campaigned vigorously against the Obama administration's Keystone Pipeline project, which was designed to transport eight hundred thousand barrels of crude oil a day from Canadian oil sands to Gulf of Mexico ports. He was arrested twice while leading anti-Keystone protests at the White House. In 2015, after seven years of promoting the pipeline, Obama reluctantly bowed to public pressure and canceled plans to build it.

The Keystone project was revived by Trump, however, almost immediately after he took office in January 2017. Trump returned to Bush's crude censorship playbook with a vengeance: His incoming administration "issued *de facto* gag orders to government science agencies like the EPA and USDA, ordered that the EPA take down its climate webpage," and "mandated that any studies or data from EPA scientists must undergo review by political appointees before they can be released to the public."[24] In response, scientists in various parts of the United States and Canada initiated "guerrilla archiving" projects, rapidly copying decades of climate records and other government data to independent servers in order to prevent their sudden disappearance.

Obama versus Trump?

The Trump administration's house of horrors does not justify distorting the record of his predecessor. Barack Obama was a far more sophisticated but no less faithful servant of the coal and oil industries. As environmental advocates were urging him to leave the fossil fuels in the ground, Obama was reassuring energy executives of his intention to do the opposite: "As long as I'm president, we are going to keep on encouraging oil development and infrastructure."[25]

To drive the point home, he claimed that the United States had produced more oil during his first four years in office than during G. W. Bush's eight-year tenure. He took credit for adding "enough new oil and gas pipeline to encircle the earth and then some." He even boasted of making more public land available for private fossil fuel exploitation. "We're drilling all over the place," he said. "And you have my word that we will keep drilling everywhere we can."

And what did Obama learn from the historic Deepwater Horizon disaster, in which British Petroleum spewed two hundred million gallons of oil into the Gulf of Mexico over an eighty-seven-day period in 2010? Less than two years later, he declared:

> Offshore, I've directed my administration to open up more than 75 percent of our potential oil resources. And that includes an area in the Gulf of Mexico that we opened up a few months ago that could produce more than 400 million barrels of oil—about 38 million acres in the gulf.[26]

Scientific Consensus versus Denial and Disinformation

Scientific consensus is not easy to quantify, but a comprehensive "study of studies"—taking into account 24,210 peer-reviewed articles on climate change—found that 99.99 percent of their 69,406 authors supported the propositions that global warming is real and primarily results from human activity, especially the burning of carbon-based fuels. With only four of 69,406 scientists dissenting, the degree of agreement is impressive, to say the least.[27]

The science-denial disinformation deliberately confuses superficial phenomena (weather) and underlying reality (climate). Abnormally cold winter weather is cited as evidence against climate change. Donald Trump expressed this line of reasoning in his inimitable style: "This very expensive GLOBAL WARMING bullshit has got to stop. Our planet is freezing, record low temps, and our GW scientists are stuck in ice."[28] His simple-minded appeal to common sense does not rise to the level of scientific argument, but because it came from the mouth (or Twitter feed) of the future occupant of the White House, it could not be ignored. Its fallacy lies in substituting a momentary sample of the earth's temperature at a part of its surface with the long-range temperature of the planet as a whole. Abnormally cold winters are, in fact, a predictable consequence of the climate instability caused by global warming.

A related fallacy attempts to devalue the climate scientists' warnings that failure to reduce CO_2 emissions will result, over the next few decades, in a rise in average global temperature of three or four degrees Fahrenheit. The denialist response is, "So what? Temperature fluctuations of twenty, thirty, or forty degrees in a single day are not unusual. Why should we worry about a paltry three or four more?" But those three or four extra degrees would in fact have catastrophic consequences, the most evident of which would be enough melting of polar ice to cause the oceans to rise and submerge some of the world's most populous cities. "The last time the world was 2 degrees warmer," Hansen explains, "sea level was 6 meters or 20 feet higher."[29]

The earth's temperature is analogous not to the air temperature you feel on your skin but to the one inside your body. If your body temperature goes up three or four degrees and does not come back down, it means you are dying. The same goes for Mother Earth.

Doubling Down on Denial: What's at Stake?

The growing public awareness of the environmental danger means that elected policymakers are increasingly defying the will of their constituents. That became more evident and explicit than ever in early 2017 when Donald Trump appointed notorious climate-change denialists to head agencies charged with defending the environment.[30] His choice for director of the Environmental Protection Agency, Scott Pruitt, had sued the EPA fourteen times in his previous capacity as attorney general of Oklahoma.[31] Pruitt's scandal-ridden tenure at the EPA lasted less than a year and a half; his replacement was Andrew Wheeler, a former coal industry lobbyist.

Trump's tweet complaining about "very expensive GLOBAL WARMING bullshit" provides the essential clue to understanding the driving force behind denialism. Calls to take action against global warming are "expensive" because they threaten the profitability of the gargantuan energy firms. As indications of the stakes involved, the global oil market was valued at $1.7 trillion per year in 2015,[32] and the value of the global coal industry was projected to reach $1.037 trillion in 2018.[33] These trillion-dollar industries do not take lightly the rising demands to halt and reverse the amount of CO_2 in the atmosphere. While some corporations have attempted to shift a portion of their investments toward new technologies that could potentially alleviate the effects of global warming, the energy industry as a whole has doggedly resisted the pressure to clean up its act. The upshot has been an all-out assault on climate and ecological science.

The dishonest nature of Big Oil's denial campaign is starkly revealed in internal communications from the archives of ExxonMobil, the world's largest publicly traded international oil and gas company.

Those documents prove that the company covered up and denied its own global warming research.

Before James E. Hansen There Was James F. Black

In 1977, eleven years before Hansen's congressional testimony made climate change a major public issue, Exxon research scientist James F. Black told the company's top management that the world's dependence upon carbon-based energy posed a grave threat to the human race.

"There is general scientific agreement," Black told the executives, "that the most likely manner in which mankind is influencing the global climate is through carbon dioxide release from the burning of fossil fuels."[34] Exxon's initial response was commendable:

> Within months the company launched its own extraordinary research into carbon dioxide from fossil fuels and its impact on the earth. Exxon's ambitious program included both empirical CO_2 sampling and rigorous climate modeling. It assembled a brain trust that would spend more than a decade deepening the company's understanding of an environmental problem that posed an existential threat to the oil business.[35]

It is ironic, from our current perspective, that pioneering climate-change research was produced under the aegis of a major oil-producing corporation. But alas, Exxon's top management was fearful of what its scientists had discovered and switched to cover-up mode:

> Toward the end of the 1980s, Exxon curtailed its carbon dioxide research. In the decades that followed, Exxon worked instead at the forefront of climate denial. It put its muscle behind efforts to manufacture doubt about the reality of global warming its own scientists had once confirmed. It lobbied to block federal and international action to control greenhouse gas emissions. It helped to erect a vast edifice of misinformation that stands to this day.[36]

Eventually, however, Exxon's duplicity was exposed. In November 2015 the attorney general of New York announced a criminal investigation into whether the ExxonMobil corporation had lied to the public about the dangers of global warming. When Exxon's lawyers asserted the company's right to free speech as a defense, the attorney general responded, "The First Amendment, ladies and gentlemen, does not give you the right to commit fraud."[37]

The Infamous Koch Brothers

Joining ExxonMobil atop the list of denialist enablers are sibling fossil-fuel moguls Charles Koch and the late David Koch, who bankrolled climate change–denial front groups to the tune of at least $100 million between 1997 and 2015.[38] The Koch brothers' combined personal wealth, estimated at around $82 billion in 2016, is derived from their control of Koch Industries, the second-largest privately held corporation in the United States.[39] Koch Industries is an oil, gas, and manufacturing conglomerate with annual revenue close to Microsoft's.

An analysis of "all known organizations and individuals promoting contrarian viewpoints . . . about climate change from 1993–2013" identified ExxonMobil and the Koch brothers as the leading financial backers of climate change denial. The Union of Concerned Scientists published the following list showing the documentable extent of their support.[40]

American Enterprise Institute
Exxon: $3,615,000, 1998–2012

Koch: >$1 million, 2004–2011

Americans For Prosperity
Koch: $3,609,281, 2007–2011

American Legislative Exchange Council
Exxon: >$1.6 million, 1998–2012

Koch: >$850,000, 1997–2011

Beacon Hill Institute at Suffolk University
Koch: ~$725,000, 2008–2011

Cato Institute (before 1976, the Charles Koch Foundation)
Koch: >$5 million, 1997–2011

Competitive Enterprise Institute
Exxon: ~$2 million, 1995–2005

Koch: $709,725, 1997–2014 [data from Greenpeace, 2014]

Heartland Institute[41]
Exxon: >$675,000, 1997–2006

Koch: $millions via Koch-funded organization Donors Trust, through 2011

Heritage Foundation
Exxon: $780,000, 2001–2012

Koch: >$4.5 million, 1997–2011

Institute for Energy Research
Exxon: $307,000, 2003–2007

Koch: $125,000, 2007–2011

Manhattan Institute for Policy Research
Exxon: $635,000 since 1998

Koch: ~$2 million, 1997–2011

The Koch brothers are emblematic of an ideological shift that put science directly in the crosshairs of disinformation. Their father, Fred Koch, had been a board member and leading financial backer of the racist, ultra-anticommunist John Birch Society, an early source of the ideological poison that has since spread widely throughout the American body politic. Fred Koch promoted the idea that desegregation was a communist plot to "enslave both the white and black man."[42]

In his youth, Charles Koch had followed in his father's John Birch footsteps, but he eventually shifted his loyalties to Ayn Randian libertarianism.[43] Rand's absolute antipathy to governmental regulation of business affairs provided the Koch brothers with a worldview in close harmony with their material interests. They considered government oversight of their private business operations an intolerable abridgement of their freedom—their *liberty*!—to manage their companies in any way they saw fit. Why, they demanded, should government agencies be empowered to stop their energy firms from using the nation's rivers and atmosphere as dumping grounds for noxious waste products? Environmental regulation, they contended, was but another form of Marxist totalitarianism. *Environmentalism was the new communism.*

Ecosocialism

The notion that environmentalism is somehow linked with socialism and anticapitalism is not entirely untrue, but it is a half truth, and, as Benjamin Franklin observed, "Half a truth is often a great lie."

The true part is that a growing number of environmental activists have begun to identify themselves as "ecosocialists" and to recognize the capitalist economic system as the number one obstacle to preserving humanity's habitat. "The future will be ecosocialist," declares author Joel Kovel, "because without ecosocialism there will be no future."[44] But the right-wing narrative—that Greens are really just Reds in disguise trying to dupe the public—has it backward. Socialists didn't create environmentalism; the environmental crisis has created socialists.[45]

As thoughtful people have awakened to the threat of ecological devastation and have sought to understand its causes, many have come to see the capitalist economic system as the prime suspect. "There is still time to avoid catastrophic warming," Naomi Klein proclaimed in a bestseller bearing the subtitle *Capitalism vs. the Climate*, "but not within the rules of capitalism as they are currently constructed. Which is surely the best argument there has ever been for changing those rules."[46]

Because those rules are indispensable to maintaining the wealth and power of ExxonMobil, the Kochs, and the current energy industry in general, any attempt to change them will of course meet with ferocious opposition. It should not surprise anyone that capitalists would resist the rise of anticapitalist political sentiment. The environmental movement, however, had achieved considerable size and strength well before ecosocialism came on the scene.

Meanwhile, the Koch agenda has attained immense influence in the American political arena. The brothers first used their financial clout to gain control of a small political entity, the Libertarian Party. David Koch was its vice presidential candidate in the 1980 elections. As their political ambitions outgrew the possibilities offered by third-party challenges, they shifted their efforts toward gaining direct influence within the Grand Old Party itself, with resounding success. An $8 million investment to create an outfit called Citizens for a Sound Economy produced faux grassroots ("astroturf") movements that presented as populist groups, fueled by outrage, but fully funded by the Kochs and others. This strategy culminated in the Tea Party, which, in 2010, eventually drove the Republican Party as a whole solidly into the Kochs' orbit.[47]

In the 2012 and 2014 elections the Kochs spent more than half a billion dollars backing Republican candidates, and their spending increased to an estimated $900 million in the 2016 campaign season.[48] Science denialists serving corporate interests thus captured one of the two major American political parties and effectively neutralized the other.

Oversight and Regulation: The EPA

The ferocity of the Trump administration's assault on the EPA understandably makes everyone concerned for the environment want to rise to the agency's defense. Nonetheless, we are obliged to consider how well the EPA has done its job. Has it in fact protected the environment in a meaningful sense? And how has science figured in the process?

When the Toxic Substance Control Act became law in 1976, the EPA was charged with administering it. The 62,000 chemicals already in production in the United States were simply given a free pass. Since then, the EPA "has approved another 22,000 chemicals, most with little or no data on the risks to health or the environment. Thus, there are more than 80,000 chemicals in use today in the U.S., most of which have not been independently tested for safety."[49] On the other side of the ledger, how many industrial chemicals have been limited or banned by EPA regulations in those four and a half decades? A grand total of *five*.[50] A report by the Government Accountability Office pointed to a primary problem with the EPA's oversight of toxic chemicals. Noting that the agency's assessment of dioxin—the active ingredient in Agent Orange linked to many cancers and birth defects—"has been under way for 18 years" and was still "years away" from completion, the GAO declared:

> The Environmental Protection Agency (EPA) lacks adequate scientific information on the toxicity of many chemicals that may be found in the environment—as well as on tens of thousands of chemicals used commercially in the United States.[51]

The EPA's deficiencies are only the beginning of the weakness of US environmental regulation. Even on the rare occasions when the EPA does impose a regulation the polluting industries don't like, the latter can appeal to a higher authority most Americans have never heard of: OIRA—the Office of Information and Regulatory Affairs.

OIRA's reversal of the EPA's decisions has become commonplace in recent years. According to a former EPA attorney, "A rule will go through years of scientific reviews and cost-benefit analyses, and then at the final stage it doesn't pass." This, she added, "has a terrible, demoralizing effect on the culture at the E.P.A."[52] OIRA is officially "within the Executive Office of the President." In other words, the president controls it. OIRA has the power to review and alter all actions taken not only by the EPA but also by the FDA, OSHA, and other regulatory agencies. Corporations that contribute large sums to presidential campaigns thus have a friend in the White House as their ultimate bulwark against unwelcome regulation.

Between 2001 and 2011, more than six thousand draft regulations were subjected to OIRA's highly secretive review process. Of those submitted to the Bush-Cheney White House, 64 percent were overturned or altered. An even higher percentage of those submitted to Barack Obama's OIRA—76 percent—were overturned or altered. "The sad reality," one disappointed Obama supporter lamented, is that the president repeatedly undermined environmental, health, and safety regulation "by allowing OIRA to substitute its judgment for the expertise of the agencies."[53]

For past administrations, both Republican and Democratic, the EPA served as a cover, allowing them to pretend that they were doing something to protect the environment. The Trump administration seems willing to forego that pretense. To the extent that the EPA is dismantled by Trump, there will be less need for an OIRA to overrule it. OIRA is not scheduled for elimination, however; instead, it will be directed to collaborate with the EPA in implementing the Trump administration's "regulatory overhaul."[54]

The Greenwashing of Corporate America

Early on, when the attack on Rachel Carson backfired, CEOs and their boards of directors learned a valuable lesson: public concern for the natural environment was a genie that could not be stuffed back in its bottle. That made a two-pronged strategy necessary: in addition to financing the disinformation campaigns, polluting corporations would also strive to co-opt the burgeoning environmentalist movement.

Our vocabulary was enriched with yet another new word—"green-washing"—defined as a public relations ploy aimed at creating the perception that a company's products or policies are "green," in other words, good for the environment. To that end, environmental scientists were hired, the language of ecology was adopted by PR departments, and green credentials were trumpeted. Corporations' new green orientation went beyond PR initiatives and into direct engagement with the sciences. Their scientists and engineers were hired primarily not to front for false claims but to perform legitimate research.

The intent was to channel environmentalism into respectable pathways that would present fewer challenges to corporate profitability. To that end, they sought to bring the movement under the control of professional environmentalists who could propose engineering solutions—rather than political solutions—to environmental problems. The technocratic approach, if successful, would minimize the impact of regulation and other forms of governmental intervention. But above all it would provide a counterweight to the movement's growing radical, anticapitalist, ecosocialist tendencies.

The new "mainstream" environmental sciences would embrace ecology as a guiding principle.

> But it was not a subversive ecology that questioned fundamental values of economics, consumer habits, and techno-scientific control. It represented an engineering mentality in which problems of waste, pollution, population, biodiversity and the toxic environment could be solved scientifically.[55]

Proposed technocratic fixes for our damaged ecosphere have been legion. They include technologies such as coal scrubbers, carbon capture and sequestration, biofuels, photovoltaic cells, wind turbines, and nuclear power, as well as economic innovations like cap-and-trade and other carbon emissions trading schemes. One of the most prominent of the suggested alternatives to dirty fossil fuels is itself a fossil fuel: natural gas.

Natural Gas as Clean Energy?
The depletion of Big Oil's crude oil reserves, combined with new natural gas extraction technologies, led the industry to begin producing less oil

and more natural gas. And as natural gas gained in importance as a "profit center," the monopolizing tendencies of Big Oil came ever more to the fore. As recently as 2009, "more than 80 percent of U.S. gas supplies were produced by 'mom-and-pop businesses'—companies with an average of a dozen employees and a market capitalization of less than $500 million."[56]

But in late 2010 ExxonMobil bought natural gas producer XTO Energy, and the industry would never be the same. Not wanting to be left behind, British Petroleum, Shell, ConocoPhillips, and Chevron quickly got into the act. Within five years, Big Oil completely dominated natural gas production.

Although Big Oil's shift toward natural gas was carried out for business reasons, it was greenwashed as a virtuous turn to producing "clean energy." Natural gas was promoted as a less-polluting alternative to coal as the primary electricity-generating fuel. As fossil fuels go, natural gas is presumed to be *relatively* clean because burning it puts only about half as many billions of tons of CO_2 a year into the atmosphere as does burning coal. That is still an unacceptable amount. Being not *as* dirty does not make it clean. But the greenwashers portray natural gas as a "bridge fuel"—one whose use can provide a bridge to a future devoid of fossil fuels altogether. That claim was echoed by Barack Obama in his 2014 State of the Union address when he touted natural gas as "the bridge fuel that can power our economy with less of the carbon pollution that causes climate change." Unfortunately, the Sierra Club and some other environmentalist organizations supported this notion.[57]

Methane: Leaking, Venting, and Flaring

The paeans to natural gas, however, take only CO_2 into consideration. Natural gas consists primarily of methane (CH_4), which "can warm the planet more than 80 times as much as the same amount of carbon dioxide."[58] It is important to understand that not only does *burning* natural gas produce greenhouse gases, it is *itself* a greenhouse gas. That would not matter if natural gas were prevented from entering the atmosphere. The energy industry assures us that their technology can accomplish that, but independent assessments say otherwise.[59] An estimated thirteen million metric tons of natural gas, enough to fuel ten million homes, leaks into the atmosphere every year as it is extracted, transported, stored, processed, and distributed.[60]

What is worse is that the oil companies deliberately release tremendous amounts of methane directly into the atmosphere in a common practice called *venting*. Added to that is another purely wasteful practice called *flaring*—"the intentional burning of natural gas as companies drill faster than pipelines can move the energy away." According to World Bank estimates, "Flaring last year emitted more than 350 million tons of carbon dioxide globally, equivalent to the greenhouse gas emissions of almost 75 million cars." In 2018 operators in the three largest US shale-oil fields "flared or vented a record 320 million cubic feet of gas, more than 40 percent above levels seen just five years ago. The pace for the first two quarters of 2019 has been even higher."[61]

In absolute terms, the worst offenders are the largest oil companies, such as Exxon and British Petroleum. But many smaller oil producers vent or flare *100 percent* of the methane their wells pump out because they see it as an unwanted by-product. One Texas shale-oil producer, Exco Resources, flares away almost all of the natural gas it produces "even though a pipeline already exists to move it away, because it is cheaper to release the gas than pay the fees to pipe it off and sell it."[62]

At the end of August 2019, the potential environmental threat from methane increased dramatically when the Trump administration announced the dismantling of existing methane regulations.[63]

Fracking

Natural gas, furthermore, has another enormous environmental downside. The technology that makes its extraction from oil shale economically attractive to the energy industry is hydraulic fracturing, better known as "fracking." Between 2000 and 2015, the share of American natural gas production due to fracking rose dramatically, from less than 5 percent to 67 percent.[64] Over the same period, the proportion of crude oil produced in the United States by fracking went from 2 percent to more than half.[65] And the Trump administration touched off a fracking frenzy by auctioning off drilling rights on millions of acres of public lands.

Fracking consists of injecting vast quantities of "frack fluid"—a mixture of water, sand, and chemicals—into subterranean shale rock at extreme pressure in order to cause cracks in the rocks, allowing trapped natural gas or oil to flow. The volume of water involved in the process is almost beyond comprehension—from two hundred to two thousand tanker truckloads *per well*.[66] As of 2016—*before* the Trump escalation—there were three hundred thousand fracking wells in the United States.[67]

The chemicals used in the fracking process are toxic, and the frack fluid is further contaminated by heavy metals and radioactive elements it picks up from the shale. Typical contaminants include benzene, toluene, ethyl benzene, xylene, formaldehyde, arsenic, cadmium, and lead. At the end of the process, industry representatives claim, this tainted liquid is recovered and disposed of safely. Independent observers, however, report that a significant amount finds its way into local rivers, streams, and groundwater, posing a serious health hazard to nearby communities. Infertility, miscarriage, and birth defects are a few of the reproductive and developmental disorders linked to frack-fluid pollutants.

Anthropogenic Earthquakes: "Human-Induced Seismicity"

An especially dramatic indication of the environmental danger posed by fracking is its tendency to physically tear the earth apart. Fracking literally causes earthquakes.

Earthquakes had always been legally defined, for insurance and other purposes, as "acts of God." Not anymore. In 2016 the US Geological Survey added a new category to its maps: "human-induced seismicity." Seven million Americans, the USGS reported, live and work in areas susceptible to earthquakes caused by wastewater disposal from oil and natural gas drilling. Most at risk are residents of Oklahoma, Kansas, Texas, Colorado, New Mexico, and Arkansas. In some cases, the chance of damage "is similar to that of natural earthquakes in high-hazard areas of California."[68] This is not an exclusively American phenomenon. Deadly earthquakes triggered by fracking and other forms of natural gas extraction have also afflicted vast territories from Alberta, Canada, to China's Sichuan province, to the Netherlands.[69]

The American epicenter, so to speak, of this unnatural seismic activity lies in central Oklahoma. In 2015 the state's government belatedly acknowledged the cause-and-effect relationship between fracking and earthquakes. The evidence had become impossible to deny: That year more than a thousand quakes of at least 3.0 magnitude occurred in the state—up from two in 2008.[70] On September 3, 2016, an earthquake measuring 5.6 on the Richter scale rattled Oklahoma and six neighboring states. State regulators immediately ordered the shutdown of thirty-seven fracking waste disposal wells over a 725-square-mile area.

Not a Bridge but a Barrier?

Natural gas will be, critics fear, not a bridge to a better environmental future but a barrier. The billions of dollars invested in natural gas technologies like fracking and horizontal drilling represent a massive misallocation of scientific resources. Those are billions that could have been, but were not, invested in alternative, noncarbon-based energy sources that hold the key to averting the looming global warming catastrophe.

The technocratic approach to solving environmental problems is not harmful in itself. If the scientists and engineers were free to set their own research agendas, they undoubtedly could make rapid strides toward lowering the CO_2 content of the atmosphere to safe and sustainable levels. But the energy corporations who hold the purse strings divert the world's scientific time and talent toward keeping the fossil fuel zombie on its feet.

No serious environmentalist, to my knowledge, denies that the earth's more than seven and a half billion inhabitants need energy, and plenty of it. Oil, coal, and gas cannot simply be dispensed with; they must be replaced with nonpolluting alternatives. Three leading candidates for that job are solar, wind, and nuclear power.

Energy from the Sun and the Wind

Wind farms and solar panels went from novelties in the last quarter of the twentieth century to familiar parts of the landscape in the twenty-first. By 2016 wind turbines were producing more than 5 percent of the electricity in the United States. Five states—Kansas, Iowa, Oklahoma, and North and South Dakota—were by then deriving more than 20 percent of their electricity from wind power. For Kansas, that percentage rose from less than 1 percent a decade earlier to almost 30 percent.

The sun's rays are a virtually unlimited source of energy. They deliver 173,000 terawatts a year to the earth's surface; our global energy consumption is a comparatively paltry 13 terawatts.[71] "More energy from sunlight strikes the Earth in one hour . . . than all the energy consumed on the planet in a year." Solar panels covering less than 0.2 percent of the earth's surface could produce double the amount of power that fossil fuels now provide.[72]

The technology for tapping this source had been in development for four decades when, in 2005, the US Department of Energy declared, "The scientific stage is set for rapid progress in solar energy research."[73] So why, more than ten years later, was solar energy still accounting for only about 1.5 percent of American electricity generation?[74]

The constant mantra of Big Oil and Big Coal is that solar energy is simply too expensive—that fossil fuels provide kilowatt-hours of energy at far lower prices. While that once may have seemed like a reasonable argument, the "current state of climate economics," economist Paul Krugman says, "has changed far more in recent years than most people seem to realize."[75] The technological progress predicted by the Department of Energy has lowered the costs of producing energy from the sun and wind to the point where they could compete in the market with fossil fuels:

> The numbers are really stunning. According to a recent report by the investment firm Lazard, the cost of electricity generation using wind power fell 61 percent from 2009 to 2015, while the cost of solar power fell 82 percent.[76]

The longer-range change has been even more striking. Bloomberg financial news reported that the cost of solar power in 2016 was 1/150th of its 1970 level, and over the same period the number of solar installations had multiplied by 115,000. Meanwhile, "clean energy investment broke new records in 2015 and is now seeing twice as much global funding as fossil fuels."[77]

What these figures suggest is that the greatest barrier to confronting climate change is not scientific, technological, or even economic, but political.

The Social Cost of Carbon Dioxide

Big Oil and Coal depend upon "corporate welfare"—massive government subsidies—to keep the industry afloat. By one credible estimate, "the total annual value of all known U.S. state and federal fossil fuel exploration, production, and consumption subsidies is $32.8 billion."[78] A report by the International Institute for Sustainable Development estimated that a worldwide transition to clean energy could be paid for with just 30 percent of the amount countries currently spend to subsidize coal, oil, and gas companies.[79]

Even the $32.8 billion-a-year estimate, enormous as it is, vastly understates the artificial advantage accorded to fossil fuels in the American political economy because it does not include the damage wrought by fossil fuels, which we also subsidize. The greatest of these hidden subsidies, the Social Cost of Carbon Dioxide, is defined by the National Academies of Sciences as

> an economic metric that provides a comprehensive estimate of the monetized value of the net damages from global climate change that results from a one-metric-ton increase in CO_2 emissions.[80]

The Social Cost of Carbon Dioxide was calculated for the year 2015 to be $36 per ton of CO_2. Multiply that by 5.4 billion tons in US emissions for that year, and you get "the $200 billion fossil fuel subsidy you've never heard of":

> This $200 billion is a cost in real money—in lost labor productivity, healthcare costs, increased energy expenditures, coastal damages—that is paid somewhere in the world for each ton of carbon dioxide that is emitted.[81]

What this means is that to get the earth back on track toward a future with an atmosphere in sustainable CO_2 balance would require not only canceling the visible $32.8 billion a year giveaway to the fossil fuel industries but also forcing them to pay the $200 billion social cost of their polluting activities as well.

The American Public Is Worried

A large majority of Americans are concerned about the threat of an impending ecological crisis. A 2016 Gallup poll reported: "Sixty-four percent of U.S. adults say they are worried a 'great deal' or 'fair amount' about global warming." It also revealed that 65 percent believe "increases in the earth's temperature over the last century are primarily attributable to human activities rather than natural causes."[82] And yet this majority has been politically marginalized. The governmental officials charged with the authority to do something to prevent the crisis either do nothing or make matters worse.

This is the clearest indication that the tragedy of American science is to a great extent a tragedy of American science *policy*. Those who hold the reins of political power in the United States have acted in the most irresponsible way imaginable: they are not only fiddling while the earth burns, they are pouring more fossil fuels on the fire. In legal parlance, their behavior qualifies as "depraved indifference."

CHAPTER 7

Atoms for Peace?

IN 1953, JUST EIGHT YEARS after the first atomic bombs demolished Hiroshima and Nagasaki, President Eisenhower announced an "Atoms for Peace" initiative. He was not promoting nuclear power as an environmentally sound alternative to dirty fossil fuel energy. Such concerns were not then high on the public agenda.

His stated goal was a worldwide scientific research program that would produce a new, inexhaustible source of inexpensive energy. "Who can doubt," he asked rhetorically,

> if the entire body of the world's scientists and engineers had adequate amounts of fissionable material with which to test and develop their ideas, that this capability would rapidly be transformed into universal, efficient, and economic usage?[1]

That goal seemed eminently reasonable at the time. How could anyone object? Many of the scientists who championed it no doubt believed they were endorsing a noble cause. Its architects, however, were only tangentially interested in peaceful uses of nuclear energy.

Atoms for Peace, launched in the context of the Cold War, was in reality a major step on the road toward the militarization of American science. Civilian atomic energy was joined at the hip to the nuclear weapons program. The two industries operated in parallel and nourished each other. There were no clear boundaries separating the nuclear research of civilian and military laboratories. Plans for commercial reactors were simply knockoffs of reactors already being developed to produce tritium

97

for bombs. Historian Peter Kuznick succinctly exposes the program's hidden motive:

> Under the cover of the peaceful atom, Eisenhower pursued the most rapid and reckless nuclear escalation in history. The US arsenal went from a little more than 1,000 nuclear weapons when he took office to approximately 22,000 when he left. But even that figure is misleading. Procurements authorized by Eisenhower continued into the 1960s, making him responsible for the levels reached during the Kennedy administration—more than 30,000 nuclear weapons. In terms of pure megatonnage, the United States amassed the equivalent of 1,360,000 Hiroshima bombs in 1961.[2]

Peaceful Nuclear Explosions?

In the early years of Atoms for Peace, the US Atomic Energy Commission "aggressively marketed nuclear power as a magic elixir that would power vehicles, feed the hungry, light the cities, heal the sick, and excavate the planet."[3] Lighting the cities might be achieved by nuclear-fueled turbines quietly generating electricity, but excavating the planet would require something much noisier: peaceful nuclear explosions, or PNEs.

In 1958 the AEC launched Operation Plowshare to explore the incongruous notion that this apocalyptic power could be harnessed for civil engineering purposes, such as excavating canals or harbors, blasting through mountains to dig highway and railroad cuts, fracturing oil shale, or strip-mining on a grand scale. One Plowshare proposal—fortunately, never acted upon—called for exploding five hydrogen bombs to dig a harbor at Cape Thompson, Alaska.

As one commentator noted, "It hardly requires an advanced degree in nuclear physics to recognize that any device small enough to fit in an aircraft yet powerful enough to level a city has serious military implications."[4] There were a number of reasons to suspect martial motives behind this pacifistic initiative. For one thing, AEC chairman Lewis Strauss acknowledged that the real purpose of Operation Plowshare was to "highlight the peaceful applications of nuclear explosive devices *and thereby create a climate of world opinion that is more favorable to weapons development and tests.*"[5] But perhaps most telling was Edward Teller's key role in initiating and promoting the program.

Teller was director of the Lawrence Livermore National Laboratory and the American science establishment's most notorious warhawk. He wanted to design and test nuclear explosives. His ambitions were frustrated, however, by restrictions on nuclear testing imposed by international agreements, so he sought ways to circumvent them. He saw PNE detonations, because they would be indistinguishable from any other nuclear explosions, as a way to camouflage military tests. A classified government study noted: "It would be very simple to conduct weapons tests under the guise of Plowshare."[6] The AEC detonated thirty-five nuclear explosions under the Plowshare aegis between December 1961 and May 1973.

Teller also sought to defuse public fears of radioactive fallout from nuclear testing by asserting that PNE tests would be clean and safe. He and three like-minded physicists published a book to lend the authority of science to that spurious claim.[7] This was a pioneering step toward the methodology of science disinformation used, in turn, by the tobacco industry.

Although PNEs for civil engineering projects proved to be a dead end, civilian nuclear power gained real-world status in the form of nuclear reactors to boil water to make steam to turn turbines to generate electricity. This expensive, ultracomplicated means to a routine end embodied the spirit of Rube Goldberg: Why do something the easy way when a far more complex way will do?

Commercial nuclear power made its debut in the United States in 1957. As of 2016, there were sixty-one commercial nuclear plants with ninety-nine reactors, operating in thirty states. Although they did not create carbon emissions, the electricity they generated was not "clean energy." The radiation produced by their radioactive fuel is far more dangerous than carbon dioxide.

From "Ban the Bomb" to "No Nukes"

Between 1957 and 2016, the fortunes of the US nuclear industry waxed and waned and waxed again. In the 1950s, the Campaign for Nuclear Disarmament in Great Britain generated antinuclear sentiment that spread to the United States. The civil rights and anti–Vietnam War movements of the 1960s gave rise to a broad social radicalization that produced a new environmental consciousness and a growing public antipathy to all things nuclear. A number of nuclear accidents and near accidents, culminating in the reactor meltdown at Three Mile Island in 1979, precipitated an antinuclear movement in the United States too massive for politicians to ignore. The nuclear industry was put on the defensive. Over the course of the 1980s, nuclear plants began to decrease in numbers.

In the early 1990s, opinion polls showed almost two-thirds of the American public opposed the construction of new commercial reactors. In January 1993 the Clinton administration came into office proclaiming opposition to nuclear expansion. The nuclear industry responded with a $20-million-a-year public relations effort to spin their product as a clean alternative to dirty coal and oil. The campaign was apparently successful in neutralizing antinuclear sentiment, and by the start of the twenty-first century a "nuclear renaissance" was underway.

Nuclear Power as the Planet's Savior?

Industry greenwashing put forward a technocratic vision of nuclear fission as a plentiful source of safe, clean, cheap energy. A number of impressively credentialed climate scientists, headed by James E. Hansen, backed the industry's claims.[8] Because Hansen had earned a great deal of public credibility as a pioneering champion of global warming awareness, his advocacy could not be summarily dismissed. Nonetheless, the industry's financial support gave his pro-nuclear pronouncements far more prominence than they could have gained on their own merits.

A major premise of Hansen's argument was that wind, solar, and hydroelectric cannot provide enough energy to power the modern world. Other climate scientists—most notably Mark Jacobson of Stanford University—insisted that those alternatives can indeed meet that challenge. Jacobson backed that contention with detailed calculations of "just how many hydroelectric dams, wave-energy systems, wind turbines, solar power plants and rooftop photovoltaic installations the world would need to run itself completely on renewable energy."[9] Hansen's appeal to environmentalist organizations to rethink their opposition to nuclear power was unsuccessful. Greenpeace, Friends of the Earth, the Sierra Club, the Natural Resources Defense Council, and some three hundred other groups rejected it.[10]

Is Nuclear Power Safe Energy?

The debate over whether wind, hydro, and solar power are adequate alternatives only indirectly addressed the make-or-break issue for nuclear power: safety. If it could be established that nuclear reactors would no longer pose routine risks (leaking radioactivity into the earth's atmosphere and waters) or catastrophic risks (spectacular meltdowns), then opposition to nuclear power would quickly disappear.

The prediction of future accidents is by its nature a speculative undertaking. The available empirical evidence derives from past occurrences. Two words make it impossible for independent arbiters to discount future catastrophic risks: Chernobyl and Fukushima. On April 26, 1986, a reactor at the Chernobyl Nuclear Power Plant in Ukraine, then part of the Soviet Union, exploded. The radiation released by the accident was ten times that of the Hiroshima bomb. The blast sent fifty tons of nuclear fuel into the atmosphere in the form of a plume of radioactive gas and dust that winds carried across central Europe and Scandinavia. About 70 percent of Chernobyl's nuclear fallout landed in a neighboring country, Belarus, contaminating about 20 percent of Belarus's agricultural land. A nineteen-mile radius around the Chernobyl plant—more than a thousand square miles—was designated as a Nuclear Exclusion Zone, from which hundreds of thousands of people were evacuated and forced to abandon their homes. Permanent ghost towns were thus created, the largest of them Pripyat, Ukraine.

How many people will fall victim to cancer due to the long-term effects of Chernobyl's radiation? That has been a contentious issue, with estimates ranging from a few thousand to tens of thousands. One United Nations agency, the International Atomic Energy Association, predicted four thousand cancer deaths, while another, the International Agency for Research on Cancer, forecast forty-one thousand additional cases of thyroid and other cancers by 2065.[11] The lower estimate, however, cannot be trusted, because the agency producing it is not independent of the nuclear industry. "The IAEA," nuclear engineer Alexander Sich explained in 1996, "is in the business of promoting nuclear energy, not discouraging it. For ten years the agency has attempted to downplay the consequences of the accident" at Chernobyl.[12] And it continues to do so to this day.

The Chernobyl disaster was a major setback for the nuclear industry's safety pretensions. Its response was to argue that because the

Chernobyl reactor design was already obsolete, a nuclear accident of this magnitude could never happen again.

And then it did. At Fukushima, Japan, on March 11, 2011, a chain of events culminated in another nuclear meltdown. An earthquake triggered a tsunami that destroyed the Daiichi Power Plant. More than fifteen thousand people were killed at the outset; several years later, twenty-five hundred more were still officially listed as missing. The nuclear fuel in the plant's reactor cores melted down and its containment vessels exploded, releasing vast amounts of radioactivity into the atmosphere and the ocean.[13]

Two hundred and thirty square miles of land around the plant, plus another eighty square miles to the northwest, were declared exclusion zones "too radioactive for human habitation." A year and a half later, "Fukushima officials stated that 159,128 people had been evicted from the exclusion zones, losing their homes and virtually all their possessions." But the problem extended far beyond the exclusion zones. Radiation levels higher than Japan's allowable exposure rate had contaminated an area nearly the size of Connecticut—forty-five hundred square miles.[14]

The same United Nations–affiliated organization that played down the effects of the Chernobyl meltdown did the same for Fukushima. A 2015 IAEA report concluded that "no discernible increased incidence of radiation-related health effects are expected among exposed members of the public and their descendants."[15] In this case, however, the IAEA's sanguine assessment was more widely shared. By 2018, most authorities had concluded that the long-term radiation threat from Fukushima fallout was minimal. But, as one knowledgeable commentator observed, "there have been plenty of other ways to die." Social and psychological shocks caused by the meltdown produced "real and profound health consequences." Furthermore, the radioactivity threat itself will linger for decades in the vast amount of radiation still contained in the inoperative Fukushima reactors.[16]

In Japan, widespread outrage forced the shutdown of all forty-eight of the country's remaining reactors, at least for the time being. Tepco, the electric utility that operated the Fukushima plant, was "in reality, bankrupt," said a Japanese political leader, but it survived with government assistance because it was deemed "too big to fail."[17] The Fukushima

Atoms for Peace? 103

catastrophe also set back the international nuclear industry. Germany closed down eight of its oldest reactors and promised to phase out all of its nuclear facilities by 2022. Switzerland pledged to do the same by 2034. In the United States, the world's largest nuclear power producer, the industry had already been in decline for economic reasons. The Fukushima disaster accelerated that trend.

Is Nuclear Power Clean Energy?

As dramatic as the Chernobyl and Fukushima meltdowns were, the nuclear industry's day-in and day-out activities are no less dangerous. Uranium mining and processing and routine power plant operations produce an ever-increasing accumulation of radioactive waste that remains potent for hundreds of thousands of years.

The main appeal of nuclear energy is that it does not increase atmospheric CO_2. That advantage, however, does not begin to offset the threat to the environment posed by radiation poisoning. Radioactive uranium and plutonium in our air and water, a Physicians for Social Responsibility spokesperson has warned, is "a potential health hazard far greater than any plague humanity has ever experienced." Ionizing radiation causes cancer and alters reproductive genes, "resulting in an increased incidence of congenitally deformed and diseased offspring, not just in the next generation, but for the rest of time."[18]

The radioactive waste problem is, in fact, out of control. More than three decades ago, the *Congressional Quarterly* reported that a solution was then already four decades overdue:

Although there had long been broad general agreement on the need for legislation to establish a comprehensive national policy for the disposal of highly radioactive nuclear waste, no such legislation had been enacted during the nearly 40 years that nuclear waste had been generated in the United States.[19]

The amount of nuclear waste "kept in various types of temporary storage" is immense. By 1982, the report continued,

about 77 million gallons of military waste, in liquid form, was stored in steel tanks, mostly in South Carolina, Washington state and Idaho. . . . In the private sector, the nation's 82 nuclear plants burned uranium fuel to produce electricity. Highly radioactive burned (spent) fuel rods were stored in pools of water at the reactor sites, but many utilities said they were running out of storage space.

Fast forward twenty years to 2002. The US Congress finally approved a plan to build a national nuclear waste repository at Yucca Mountain, Nevada. If all went according to plan, the radioactive waste from commercial reactors and weapons production would be safely sequestered there for the next ten thousand years. Alas, the state of Nevada objected to being the national nuclear dumping ground, so the project went into limbo. The Yucca Mountain Repository is now projected to be completed by . . . 2048.[20] Until then, more than eighty thousand tons of spent fuel will presumably continue to be stored in pools or dry steel-and-concrete casks at a hundred or so separate sites across the country—the ones that had been running out of space in 1982, and that had never been designed for long-term storage in the first place.

Apologists for the nuclear industry cite scientific research to downplay the waste disposal problem and to suggest that state-of-the-art nuclear power plants pose no serious risk of radiation leakage.[21] These claims cannot be taken at face value because science controlled by commercial interests is inherently untrustworthy. As the editors of *Scientific American* note, "The industry and the US Nuclear Regulatory Commission (NRC) claim that nuclear power is safe, but their lack of transparency does not inspire confidence."[22]

Is Nuclear Power Cheap Energy?

The global nuclear industry has long promoted its product as economically viable, able to compete in the marketplace with fossil fuels. That has never been true; its competitiveness has always been entirely artificial.

The world's most successful national nuclear program, in terms of the share of its country's energy it provides, is France's. Eighty percent of France's electricity is generated by nuclear plants, which no doubt contributes to shrinking its "carbon footprint" to only about one-tenth that of the United States.[23] Price competition, however, is not a concern of the French nuclear industry because it is wholly nationalized. Its financial support comes entirely from the French government.

In the United States, by contrast, the civilian nuclear industry has been privatized from the beginning. Its technology originated in wartime research under government control, but in the postwar period the business of generating electricity by nuclear reactors was turned over to private investors. Speaking for the industry, General Electric declared in 1954 that "in five years—certainly within ten," nuclear reactors would be "privately financed, built without government subsidy."[24] More than sixty-five years later, that promise has remained unfulfilled. Government support for fledgling industries is commonplace, but they are expected to outgrow the need for subsidization. The nuclear industry, however, has never come anywhere near achieving financial independence.

A comprehensive study by the Union of Concerned Scientists confirms that "historical subsidies to nuclear power have already resulted in hundreds of billions of dollars in costs" borne by taxpayers and utility companies' customers.[25] Direct cash payouts are but a small part of the governmental support. Much more important are the tax credits, loan guarantees, and caps on liability for accidents that relieve the investors of construction costs and the burden of operating risks.

The value of federal loan guarantees is beyond measure. Without the government's promise to bail out defaulting borrowers, no private lending institution would extend credit—no matter how high the interest rate—to such a risk-plagued industry. Those risks make nuclear power plants uninsurable without federal intervention. The Price-Anderson Act of 1957, by "placing a cap, or ceiling, on the total amount of liability each nuclear plant licensee faced in the event of an accident," shifts the burden of nuclear risk from private investors to American taxpayers.[26] As a member of the California Energy Commission explained, without the liability cap, "there would be no nuclear plants."[27]

The fact that nuclear power plants cannot be privately insured exposes two fundamental industry claims as false: First, that nuclear power is an economically viable energy source. Second, that nuclear power is safe.

Cleaning Up the Nuclear Industry's Mess

The failed Yucca Mountain Repository is a monumental case in point. If the nuclear plant operators were competing on a genuinely free-market basis, they would bear the costs of safely disposing of their deadly by-products. Instead, the government has assumed that responsibility—an enormous indirect subsidy to the nuclear industry.

A law passed in 1982—the Nuclear Waste Policy Act—charged the Department of Energy with creating a national repository for radioactive waste. To pay for it, the law imposed a tax on consumers of nuclear-generated electricity, the proceeds of which were to be deposited into a Nuclear Waste Fund. From 1983 through 2013 the fund collected some $37 billion, of which $7 billion were spent on the Yucca Mountain Repository. When that project was shelved, a federal court ruled that the Energy Department had to cease

> collecting fees of about $750 million a year that are paid by consumers and intended to fund a program for the disposal of nuclear waste. The reason, the court said, is that *there is no such program.*[28]

Meanwhile, radioactive waste continues to pile up at reactor sites, forcing their owners to build ever more steel storage casks to contain it. The industry, with remarkable chutzpah, sued the Department of Energy for failing to clean up its mess. The litigation was successful; in 2014 the federal government "paid industry about $4.5 billion in damages and has projected future liabilities at about $22.6 billion."[29] All of the money spent on a nonexistent waste repository has itself been wasted. Nonetheless, it represents a mammoth subsidy to the nuclear industry at taxpayer expense. If the government had not assumed legal and financial responsibility for nuclear waste, the leaky storage sites would likely have buried the industry under a mountain of lawsuits long ago.

The Bottom Line for the Nuclear Industry

Nuclear energy is neither clean, nor safe, nor economical. It is highly unpopular and justifiably feared. The depth of the industry's financial woes was exposed in March 2017 when the Westinghouse Electric Company filed for bankruptcy. Westinghouse is owned by the mighty Toshiba corporation, which, as builder of two of the Fukushima reactors, already faced formidable difficulties.

Nevertheless, the American civilian nuclear industry will no doubt survive because the government subsidies will continue to be as generous as they have to be. A government that cuts funding for education, health care, and infrastructure due to insufficient tax revenues always finds the money to keep the nuclear industry on life support. That is a reflection of its priorities, and military considerations top the list. Although the civilian-military nuclear connection is rarely mentioned directly by US officials, it comes through loud and clear in their fears that Iranian "atoms for peace" projects could potentially produce atom bombs. Nuclear plants will endure in the United States as long as American policymakers perceive them as essential to the US atomic weapons program.

CHAPTER 8

The Academic-Industrial Complex

BETWEEN 1964 AND 1967, I WAS A WILLING yet unwitting participant in the co-optation of university science by the corporate world. Several of my professors at Georgia Tech were moonlighting as experimental psychologists in the human factors laboratory at Lockheed Aircraft. When they offered to help me get an entry-level job in the lab after I graduated, I eagerly embraced the opportunity.

I was unaware of any conflict of interest or ethical dilemma. It seemed reasonable to me (insofar as I thought of it at all) that an aircraft manufacturer required the benefits science could provide and was willing to pay for them. No one seemed to have an ax to grind; no experiments seemed tailored to fit preconceived outcomes. Where was the problem?

Many years later, with benefit of hindsight and a broader view of the world, I came to understand that the problem was the overwhelming power of corporate money to distort the priorities of American science. Lockheed functioned as a mammoth funding conduit for the Pentagon, steering billions of dollars in scientific resources toward military research. And in the course of serving their own corporate needs, Lockheed and other military contractors dragged university research facilities with them into the militarization of science.

The Rise of the Academic-Industrial Complex

But that was in the 1960s, which belong to the prehistory of the academic-industrial complex. The Lockheed laboratory, although staffed with Georgia Tech professors, was not officially affiliated with the university. A 1982 report in *Science*, the journal of the American Association for

108

the Advancement of Science, called attention to a new, emerging trend: "During the past two years, corporate investment in academic science has proliferated at major research universities throughout the United States."[1]

There was no doubt about what was driving this new relationship: "From the university's point of view, the special appeal of the burgeoning industrial connection is quite simple—money."[2] The decline in federal funding for science during the 1970s was accelerating in the 1980s due to the Reagan administration's budget-cutting agenda. But as the adage goes, when one door closes, another opens. The year Reagan was elected, the Bayh-Dole Act paved the way to corporate ownership of patents derived from taxpayer-funded research, thus transforming public knowledge into private property.[3] The financial rewards for university-corporate collaboration increased considerably for both sides. Traditional norms prohibiting professorial profiteering rapidly fell by the wayside.

The financial arrangements that had elicited the journal's alarm seem quaintly petite today. Nonetheless, a corner had been turned: "Scientists who 10 years ago would have snubbed their academic noses at industrial money now eagerly seek it out."[4] Corporations were investing in academic research, the journal continued, because they coveted the universities' research talent and technical skill. Molecular biologists were especially in demand; the commercial potential of recombinant DNA technology had made them "a commodity of considerable interest to corporations."[5] Note that in this interpretation the scientists themselves, not the technology, had been commodified.

Another manifestation of the new infusion of commercial values into academic science was the sudden appearance in the early 1980s of hundreds of small biotech companies headed by university scientists. "*A majority of the country's leading researchers in molecular genetics and related disciplines* are known to have affiliations with these new, highly competitive companies."[6] These small fish were expected to eventually serve as feedstock for bigger fish:

> Many of the major pharmaceutical and chemical houses are currently beefing up their own capacity for research in molecular biology—Upjohn, Monsanto, and Allied Chemical among them....
> These corporations can be expected to acquire or drive out some

of the small, scientist-founded biotechnology companies of which there now are nearly 200.[7]

From 1982 to the Present

The academic-industrial complex has come a long way since 1982. Persistent and deepening funding cutbacks have continued to force public and private universities alike to turn to corporations and financiers for survival. In 1982 industry was spending an estimated $200 million a year on research in American universities, only about 4 percent of the amount of federal funding received by those institutions. While corresponding figures for today are virtually impossible to calculate, industry's investment in university research is now in the billions rather than millions. And as the corporate component has ballooned, government support has steadily declined.

From the Second World War until recently, most basic science research was financed by the federal government. Corporations had traditionally put their money into applied research and left basic research to the feds, because applied research pays off in short-term profits whereas the pursuit of fundamental knowledge generally does not. Throughout the 1960s and '70s, the federal share was over 70 percent. It was still above 60 percent in 2004, but in 2013 it dropped below 50 percent and stood at 44 percent in 2015.[8]

The amount of present-day corporate financing is generally underestimated because much of it is deliberately hidden from view. For one thing, corporations can disguise their research funding as tax-deductible donations to university foundations. Although some states require the donations to be a matter of public record, many do not. Another way that the money influencing research is concealed is via Industry Affiliates programs, which for a fixed fee "provide corporate partners with a wide range of benefits, like a guaranteed seat on an advisory board, early access to student résumés, private recruiting sessions, or full access to in-person meetings on campus."[9] Virginia Tech offers thirty Industry Affiliate programs. A former director of one such program at Purdue University acknowledges that the arrangements are "controversial," because "it has an appearance, by paying a fee, of buying a researcher."[10]

Although publicly funded research is monitored by the US Office of Research Integrity, corporate-funded research has no such oversight. Corporate sponsors can therefore freely influence every stage of the process, from the experimental design to the size and focus of studies to the interpretation of their results.

Rensselaer Tech:
An "Emerging Paradigm for Higher Education"

STEM studies, encompassing science, technology, engineering, and mathematics, is a relatively new acronym for technical education. If you were asked to name well-known institutions of higher STEM learning, chances are you would include my alma mater, the Georgia Institute of Technology, on a short list headed by MIT and Cal Tech. Alas, Georgia Tech owes its name recognition mostly to a history of football prowess and its famous fight song, "I'm a ramblin' wreck from Georgia Tech and a heck of an engineer. . . ."

Rensselaer Polytechnical Institute (RPI), by contrast, has far less in the way of name recognition, but way back in the 1960s my classmates and I were envious of its sterling reputation among STEM cognoscenti. Founded in 1824, Rensselaer was the first technological research university in the United States. More recently, unfortunately, it has become best known for its high-profile role within the academic-industrial complex.

Rensselaer's president, Dr. Shirley Ann Jackson, has an exemplary résumé. She is a theoretical physicist, and among other accomplishments has served as chair of the Nuclear Regulatory Commission and on the President's Council of Advisors on Science and Technology. Dr. Jackson was featured in a 2014 *New York Times* report that identified her as "part of a cozy and lucrative club: presidents and other senior university officials who cross from academia into the business world to serve on corporate boards."[11] She sits on the boards of directors of numerous companies, including IBM, Marathon Oil, and FedEx.

The system of interlocking directorships that entwines Rensselaer with IBM and other companies integrates university research ever more thoroughly into the corporate world. By 2006 more than 50 percent of public and private universities granting doctoral degrees had presidents who were serving as corporate board members.[12]

In 2017, Dr. Jackson's corporate connections paid off in the form of a $100 million supercomputer given by IBM to Rensselaer. Also in 2017, Rensselaer was a key beneficiary of a $200 million endeavor bringing to its campus a new biopharmaceuticals institute bearing the unwieldy acronym NIIMBL. NIIMBL was created by combining a $70 million grant from the US Commerce Department with "an initial private investment of at least $129 million from a consortium of more than 150 companies, educational institutions, nonprofits, and state governments."[13]

The Ivory Tower Gives Way to the Public-Private Partnership

NIIMBL's hybrid funding makes it a prime example of the vaunted public-private partnerships, or PPPs, that have increasingly characterized the academic-industrial complex. PPPs allow private companies to leverage public funds to carry out their for-profit research agendas. Tax dollars are used to bolster their investments by building research facilities on prestigious campuses, to establish brand-name academic chairs, to fund scholarships for future scientists they hope to recruit, and so forth.

NIIMBL was but one facet of Dr. Jackson's multibillion-dollar Rensselaer Plan to construct what she calls "The New Polytechnic."[14] Rensselaer publicity reveals one way this "emerging paradigm for higher education" serves private corporate interests:

> Collaborating colleges and universities will work with industry to provide education and training programs, curriculum development, and certification standards that will ensure a pipeline of skilled workers.[15]

Dr. Jackson's vision "represents the 'tip of the iceberg' of America's trend toward a corporate model for universities," critics have charged.[16] What that model entails is Rensselaer itself becoming primarily a mammoth financial institution and only secondarily the educator of a new generation of scientists (a.k.a. the "pipeline of skilled workers").

The Iceberg of Which Rensselaer Is the Tip

Rensselaer's example is instructive but not atypical. Some universities, one astute commentator observed, "have struck bargains only Faust could love."[17] By unfortunate coincidence, one of the more important representatives of university-corporate entanglement—"the first sitting Harvard president to agree to serve on a corporate board," Dr. Drew Gilpin Faust—happened to bear that surname.[18]

But if Harvard's Dr. Faust and her peers play the title role in this version of *Doctor Faustus*, the devil's bargain is the unrelenting drive for private profit that has starved our society's public sector of resources. Education at every level has suffered mightily, and the crisis of public education has burrowed deeply into the private universities as well. Even those as enviably endowed as Harvard have been affected. In the past, Harvard's most distinguished researchers

> had no trouble winning millions of dollars in government grants to conduct their work. Now, with federal research funding flat-lining, even the best in their fields have to court new sources of support. Following their peers at other universities that have already felt the pinch, Harvard scientists are increasingly appealing to corporations and wealthy philanthropists.[19]

Microsoft, GlaxoSmithKline, the chemical giant BASF, and the Bill and Melinda Gates Foundation are but a few who have answered the Harvard researchers' pleas. Bioengineering professor Kit Parker lamented, "I've gotten to the point now that running my research group is making me less and less of a scientist; I'm a hustler."[20]

Meanwhile, on the Other Side of the Country . . .

The University of California at Berkeley, like Harvard, is an illustrious institution of higher learning with a strong desire to boost its STEM cred. But in 2013 investigative journalist Joaquin Palomino reported a familiar dilemma:

> The UC system is projecting a $2.5 billion structural shortfall by 2015, and the university's financial managers have no concrete plans to address the crisis. As a result, private money has continued to flow into Berkeley's hard science departments with little public input.[21]

Earlier, in 1997, UC Berkeley's biology department formed a public-private partnership with Novartis to conduct GMO studies. That touched off "an explosion in privately funded research."[22] In 2007, Berkeley and the University of Illinois joined British Petroleum in forming the Energy Biosciences Institute—"the largest public-private partnership of its kind in the world."[23] British Petroleum's primary interest in the $500 million venture was to further biofuels research, but it derived other valuable benefits from the deal as well.

In 2010, British Petroleum's Deepwater Horizon disaster dumped 210 million gallons of oil into the Gulf of Mexico, contaminating 665 miles of coastline.

> Just one month after the ruptured well was sealed, UC Berkeley ecologist Terry Hazen made a groundbreaking discovery: He identified a new microorganism that was eating the spilled oil and breaking it down into CO_2 and water. The microbe was so active that, according to Hazen and his team of scientists, the vast plumes of oil in the gulf had "went away fairly rapidly after the well was capped." Hazen's findings were published in the academic journal *Science*, and subsequently reported by most major media outlets. One fact, however, was often omitted: The research was funded by BP.[24]

"When people read the *Science* report, they thought they were reading a Berkeley professor's research," a professor of microbial ecology said. "They didn't realize it was also BP saying, 'You shouldn't worry about the oil spill anymore.'"[25]

Corporate cash has not only corroded the scientific integrity of UC Berkeley research but also transformed the character of the university as a whole. In 2012, Berkeley's chancellor paradoxically declared that "to guarantee our public character, we need to increase substantially our private support."[26] The president of the University of California went further, insisting that it had ceased to be a public institution and had become a "hybrid" university with more private than public characteristics. It must, he said, "adjust to this new reality."[27]

Why Does It Matter?

The ethical concerns expressed in the 1982 *Science* report are faint echoes from a remote past. The floodgates are now fully open. As the many examples cited in this and the previous chapters illustrate, the fusion of "business ethics" with academic research goes directly against the grain of scientific integrity and reliability:

- Studies produced under such conditions are designed from the start to favor sponsors' products over alternatives.

- The data collection and analysis methods are controlled by the sponsor.

- If the desired results are attained, credentialed scientists (who may or may not have been part of the investigative team) submit the study under their names to prestigious scientific journals.

- If the study cannot produce what the sponsor wants, it will not be published.

- Meanwhile, university researchers on corporate payrolls serve on federal scientific panels that regulate the industries' operations and products.

The public-private partnerships also promote the privatization of scientific knowledge, which is anathema to the ideal of science as a public asset. Scientific progress requires *open science*—maximizing the sharing of research findings as widely as possible, not restricting them.

I suspect most university scientists make honest efforts to uphold the integrity of their respective disciplines, but only the most courageous and committed can withstand the unrelenting pressures of careerism. The corporate Mephistopheles offers well-funded projects with the lure of fortune, prestige, and high positions in industry. Even the well-meaning scientists who accept corporate funding because the alternative is to give up their research entirely are often unaware—or in denial of—how their work is manipulated. Blatant attempts to directly dictate research outcomes are not the norm. Far more problematic are the ever-present subtler influences on the priorities or parameters of the studies.

More generally, the corporatization of university research "transforms students into customers, teachers into workers, administrators into CEOs, and campuses into market populations."[28] Students, it should be added, are only customers while they are in school; as we have seen, their education is designed to commodify them into "skilled workers." Rensselaer's "emerging paradigm for higher education" has apparently won the day.

CHAPTER 9

Think Tanks
and the Betrayal of Reason

As compromised as science has become within the academic-industrial complex, things could be worse. They *are* worse—far worse—outside the universities.

Think tanks—a catchy moniker for policy research institutes—are often described as "universities without students." In their ideal form, they consist of academics with no distracting teaching duties—intellectuals paid to do nothing more than read, research, deliberate, discuss, and write. Think-tank thinkers are expected to focus their brain power on the most pressing social, political, and economic problems of the day, and to propose solutions. Indeed, proposals emanating from think tanks are often heeded and implemented by political decision-makers.

But the contemporary think tank is a contradictory phenomenon. Although its credibility depends on the perception of its integrity, its survival depends on large-scale fundraising, so its reality is an unalloyed conflict of interest. A pair of investigative journalists concluded, in a comprehensive 2016 report, that think tanks

> are seen as researchers independent of moneyed interests. But in the chase for funds, think tanks are pushing agendas important to corporate donors, at times blurring the line between researchers and lobbyists. And they are doing so while reaping the benefits of their tax-exempt status, sometimes without disclosing their connections to corporate interests.[1]

The ideal that inspired the earliest think tanks was that objective, impartial examination of controversial problems by the best minds would generate optimal solutions. But because the arbiters of the proffered advice were elected officials and political appointees, the contamination of the think tanks by partisan politics was inevitable. As they evolved, individual think tanks began self-sorting into "left-leaning," "right-leaning," or "centrist" categories. The increasing polarization of American politics has thus generated a grotesque caricature of what the think-tank movement originally set out to be.

The majority of today's think tanks prioritize spin over objective research and vie for media attention by exaggerating and sensationalizing their "product." To attract funding, they tailor their research to their donors' specifications. The false debate over global warming, as a key example, was fueled by Big Energy's ability to shape the research and findings of think tanks such as the Heartland Institute and the Institute for Energy Research.

Although liberal and progressive think tanks are as compromised in this regard as their right-wing counterparts, the practice of corrupting research has been driven to extremes by billionaire libertarian ideologues like the Kochs, the DeVoses, and the Mercers. The increasingly strident presence of think tanks in American intellectual life has undermined independent research and inflicted immeasurable damage on science and society.

A Brief History of American Think Tanks

The proliferation of think tanks in post–World War II America institutionalized intellectual activity in a new and different way. The American think tank originated with the Carnegie Endowment for International Peace in 1910 and the Brookings Institution in 1916. After 1945, their numerical increase was explosive—from a few dozen to thousands throughout the world.

The devolution of the think tank from "universities without students" into barely disguised advocacy and public relations outfits began in the late 1930s. Conservative grumblings that the Brookings Institution was "too left-leaning" led to the creation of the American Enterprise Association (AEA), which later changed its name to American Enterprise Institute

(AEI). Brookings was only left-leaning, however, from a staunchly right-wing point of view. "When AEA formed in 1938," think tank historian Jason Stahl notes, "there was no doubt why it was forming—to combat New Deal liberalism with a conservative anti–welfare state position *similar to the Brookings position of the same period.*"[2]

The Second World War interrupted the process for several years, but within a year of the war's end a defense contractor, Douglas Aircraft, founded the RAND Corporation to offer policy advice to the US military. Two years later, RAND officially separated from the aircraft company to become a nominally independent research institute.[3] At that time there were about forty-five think tanks in the United States. Today there are some two thousand, with four hundred located in and around Washington, DC.

A turning point in think-tank history has been attributed to a particular incident that took place in 1971. The AEI had been asked to provide policy advice regarding a major Pentagon expenditure. Its reliably conservative proposals were duly offered, but deliberately withheld until after a legislative vote on the issue had occurred. When asked to explain the delay, the AEI president stated that they "didn't want to try to affect the outcome of the vote."[4] This annoyed some business-oriented observers: What good is the AEI's laissez-faire counsel if it doesn't influence the legislative decision-making process? Their displeasure prompted the founding of the Heritage Foundation in 1973—a new kind of think tank, devoted primarily not to studying issues but to advocacy.

Advocacy-oriented outfits on the Heritage model dominated the think-tank landscape throughout the 1980s and '90s. Then they devolved further, into "public relations think tanks" that were even less scholarly, more openly ideological, and more media-oriented.

The change in focus from research to lobbying and PR was accompanied by a steady reduction in the weight of credentialed scholars in their composition. The percentage of PhDs in think tanks declined from 53 percent before 1960 to 23 percent from 1960 to 1980, and to 13 percent after 1980.[5]

The Mont Pelerin Proliferation

The Mont Pelerin Society, an international organization of elite scholars and policymakers, played a seminal role in the postwar think-tank movement. Founded in 1947 by Friedrich Hayek and like-minded academics, the MPS has served ever since as a primary hothouse of radical libertarian economic ideology. The Mont Pelerin Society was the granddaddy of right-wing think tanks. In 1955 it created the Institute of Economic Affairs (IEA) in London, later described by a Mont Pelerin president as "Margaret Thatcher's think tank." The IEA in turn produced "a dozen think tanks (including the Atlas network), that mostly function as fronts for the MPS."[6] The Atlas Network is an umbrella organization that coordinates the activities of more than 460 think tanks and operatives in ninety-six countries.

Mont Pelerin influence in the United States was evidenced by the presence of twenty-two MPS members serving as economic advisors to Ronald Reagan during his 1980 presidential campaign. Once in office, Reagan's preferred think tank was the Heritage Foundation, whose guiding spirit was Mont Pelerin stalwart Edwin Feulner. Feulner, Heritage's president from 1977 to 2013, also served a term as president of the Mont Pelerin Society, a distinction he shared with Hayek, Milton Friedman, and James M. Buchanan.[7]

"Wingnut Welfare"

A major factor driving the metamorphosis of the twentieth-century policy research institute into the twenty-first-century think tank has been the self-interest of billionaires wanting to lower their taxes and eliminate governmental regulation of their businesses. In the process, they have opened up vast employment opportunities for professional propagandists—a state of affairs critics have lampooned as "wingnut welfare."

A full accounting of who funds which think tanks is not possible. More often than not, they and their donors prefer to downplay their relationship. Think tanks that qualify as nonprofits for tax purposes are required to disclose their funding sources to the IRS, and those disclosures are a matter of public record. Nevertheless, where donations actually come from is easy to hide. One popular method is to channel money through donor-directed funds such as Donors Trust and Donors

Capital Fund, which allow participants to legally make anonymous contributions, thus concealing their identities.

Corporate and foundation support often represents individual donors in disguise. Does it really matter whether a think tank takes its money from Charles Koch, the late David Koch, Koch Industries, or one of several "charitable" foundations the Kochs control? To wit:

- Charles Koch Foundation
- Charles G. Koch Charitable Foundation
- Charles Koch Institute
- David H. Koch Charitable Foundation
- Fred C. and Mary R. Koch Foundation
- Koch Cultural Trust
- Claude R. Lambe Foundation

The Koch brothers have been the most consequential of the think tank angels. A small sample of those graced by their largesse are the Cato Institute, the Heritage Foundation, the Heartland Institute, the American Enterprise Institute, and the Manhattan Institute. Furthermore, the Kochs have leveraged their influence by organizing a sizable network of like-minded wealthy donors.[8]

They have also donated to colleges and universities—first and foremost George Mason University, but to many, many others as well. Koch money has found its way to more than three hundred American institutions of higher learning—from Seattle Pacific University to Florida Atlantic University and from Stanford to James Sprunt Community College in Kenansville, North Carolina.[9] The brothers' unsubtle efforts at influencing course content are meeting with resistance in the form of a growing student movement called "UnKoch My Campus."

In May 2018, UnKoch My Campus released documents obtained through the Freedom of Information Act demonstrating that George Mason University had given the Koch Foundation "a voice in hiring and firing professors."[10] The university's president was forced to acknowledge that from 2003 to 2011 the university had allowed donors to take part in selecting and evaluating faculty. Tax records also revealed that the Kochs had donated $48 million to GMU from 2011 to 2014.[11]

The multiplicity of Koch funding operations (sometimes irreverently referred to as the "Kochtopus") are not the only high rollers in the right-wing think-tank game, but the names of the others are less well known to the general public. Among the most influential (in no particular order) are:

- Richard Mellon Scaife family foundations

- John M. Olin Foundation

- Lynde and Harry Bradley Foundation

- Gilder Foundation

- Walton Family Foundation

- Dick and Betsy DeVos Foundation

- Elsa and Edgar Prince Foundation

- Mercer Family Foundation

- Peter G. Peterson Foundation

- Adolph Coors Foundation

Foundations are usually referred to as charitable or philanthropic institutions, but their primary purpose for wealthy families is as tax shields. Sheltering their money in tax-exempt foundations is a way to maximize the portion of their private fortunes they can control.

A Bizarro-World Intelligentsia

The billionaires' bountiful funding created a demand for venal ideologues and an alternate academic universe in the form of a labyrinth of think tanks, foundations, donor-directed funds, and lobbying firms, all succinctly described by economist Marshall Steinbaum:

> When nothing that these institutes and individuals produce makes it into the leading journals or marquee conferences, they create their own, complete with all the trappings of academic seriousness and fancy-sounding awards, appointments, publications, and so on. This smoke and mirrors strategy serves a dual mission. First, it deceives those in the media and policy-making world into believing

that actual scholars sanctioned the work—or at least, allows them to plausibly claim this is the case. And second, it deceives its own insiders into thinking they are successful, high-status academics, not coddled cogs in a ruthless ideological-political machine.[12]

The think tanks allow politicians to claim the laws they pass are based on genuine science and scholarship. No one can credibly assert that testimony offered by think-tank spin doctors on corporate payrolls is objective. Yet politicians at all levels of government listen to it and pretend to treat it seriously, and the major news media generally go along with the charade. It is illegal for these tax-exempt groups to lobby, but that in essence is what they do.

Examinations of two think tanks from opposite ends of the credibility spectrum—the Heartland Institute and the Brookings Institution—illustrate and document how conflicts of interest and the policy institutes are inseparable.

The Think Tank That Specializes in Climate-Change Denial

Among the think tanks least likely to pass a smell test is the Heartland Institute. Satirists call it the Flatland Institute because of its aggressively anti-science agenda. The *Economist* described Heartland as "the world's most prominent thank tank promoting skepticism about the man-made climate change."[13]

In common with most think tanks, Heartland goes to great lengths to hide its funding sources and camouflage the purely partisan lobbying nature of its operations. In 2012, however, leaked emails exposed a trove of the organization's confidential internal documents to public scrutiny, providing a rare glimpse into Heartland's inner workings.[14] The emails revealed, among other things, Heartland soliciting donations to finance lavish "educational" campaigns. One focused on designing and promoting kindergarten-through-high-school curricula to inject doubt about climate science into impressionable young minds.[15] Meanwhile, Heartland was also bankrolling the Nongovernmental International Panel on Climate Change (NIPCC), to the tune of $300,000 a year. The NIPCC holds large, annual, convention-style global-warming-denial gatherings mocked by environmentalists as "denial-paloozas."

Heartland's fundraising appeals target oil companies and "other corporations whose interests are threatened by climate policies." The think tank's proposed budget (stamped "Confidential. Please Do Not Circulate.") allocated two-thirds of its $6.4 million in projected expenditures to such categories as Government Relations, Fundraising, Communications, and Publications.[16] That is the budget of a lobbying outfit, not a research institute.

Think tanks like Heartland are virtually unregulated, but they do have one very clear legal limitation. To maintain their tax-exempt status, they are required to strictly avoid partisan politics, but that is narrowly defined as directly supporting a candidate running for office. Most think tanks get around that restriction by simply setting up parallel organizations that are not tax-exempt.

The revelations of the Heartland emails were hardly shocking, because for the most part they merely confirmed what impartial observers had long assumed. Yet not all think tanks are as obviously ethically compromised as Heartland, Heritage, or Cato.

The Brookings Institution: "The Top Think Tank Worldwide"

The pioneering Brookings Institution provided the archetype for the modern think tank. At its founding, in 1916, it set a high standard of objective scholarship and remains the most prestigious think tank in the world. That is the judgment of the University of Pennsylvania's Lauder Institute, a sort of meta–think tank devoted entirely to studying and ranking think tanks.[17] The Lauder Institute's Think Tanks and Civil Societies Program maintains a database of almost seven thousand think tanks and publishes an annual "Global Go To Think Tank Index Report." Its 2016 report ranks Brookings at the head of its list of "Top Think Tanks Worldwide." Among the criteria on which the ranking is based are "the quality and reputation of the research and analysis produced."[18]

But Brookings's internal memos and confidential correspondence with its donors reveal that its reputation for independent scholarship is not entirely deserved. The documents, brought to light by the New England Center for Investigative Reporting in collaboration with the *New York Times*, show Brookings routinely soliciting financial support by promising "donation benefits" to major corporations. One of the documents, a confidential spreadsheet,

> listed nearly 90 corporations, from Alcoa to Wells Fargo, providing a glimpse of the vast electronic file that Brookings maintained on donors and prospects, and the benefits it might offer. The database, along with thousands of pages of emails, solicitations for money and memos on meetings with corporate officials, highlighted Brookings's practice of assuring that donors would see results from their contributions.[19]

Brookings's memos and emails leave no room to doubt the organization's full awareness that its research and its ideas are of secondary concern to the corporations it partners with. The quid pro quo for the donations is use of Brookings's prestige to further corporate marketing efforts.

Brookings and the Sciences

While the right-wing think tank movement attacks American science head-on, the more respectable policy institutes like Brookings represent a subtler threat. Brookings's primary areas of focus have been in the social sciences, but it has also directly involved itself in other fields as well. Consider its dealings with Big Pharma, as exemplified by the case of Dr. Mark B. McClellan, who has worn many hats in his career.

Dr. McClellan, as a Brookings Institution Senior Fellow who headed up a health-care studies program, wore the hat of a pharmacological expert who could offer scientific testimony to governmental health care panels. As an FDA commissioner, he wore the hat of a government official who made consequential health care decisions. And as a member of Johnson & Johnson's board of directors, he wore a Big Pharma hat.[20] Johnson & Johnson's hepatitis C drug earned $2.3 billion in sales revenue in its first year of production. A patient who needed that medicine would be faced with paying $66,000 for a twelve-week course of treatment. Switching to his Brookings Senior Fellow hat, Dr. McClellan promoted the efficacy of that class of hepatitis C drugs. In 2014, Dr. McClellan earned about $265,000 from Johnson & Johnson and $353,145 from the Brookings Institution. As obvious as that conflict of interest would appear, it was evidently not obvious enough to the organizations engaging Dr. McClellan.

Although the Brookings name, like those of certain elite universities, continues to command respect in the halls of power, it is no less immune to corporate influence than the execrable Heartland or Cato Institutes. Brookings is thus complicit in vitiating the integrity of American science. In recent decades, however, Brookings has been pushed to the periphery, partially eclipsed by its libertarian rivals.

Trump as Negation and Apotheosis

Given this chapter's focus on intelligence and intellectuals, it is perhaps unsurprising that the name "Trump" has thus far not appeared in it. However, Trump's rise represents both the negation and the apotheosis of the right-wing counterintelligentsia. Many public intellectuals of the right are torn between repugnance for the man himself and exuberance at the prospect that they can use him to accomplish their goal of "dismantling the administrative state."

The libertarian scholars are no doubt embarrassed by Trump's truculent ignorance, but his administration represents a magnificent opportunity to destroy public faith in "big government" by discrediting it.

Conclusion: What Think Tanks Have Done to American Science

The denizens of the think-tank movement created according to the Mont Pelerin blueprint have sold themselves for money, for career, for material comforts, for empty prestige and counterfeit honors. It is a worldwide phenomenon, but its essential core emerged in postwar America.

The whole operation has been founded on dishonesty and ignoble lies. Its purpose has been Robin Hood's in reverse—to transfer the vast wealth of the American economy from the households of the many to the coffers of the few. Already the greatest heist in human history, a robbery of trillions upon trillions of dollars, its perpetrators are not yet satisfied. And they are still at large.

And what has this to do with American science? In the process of perverting the intellectual vocation, the right-wing intelligentsia has facilitated a political assault on empirical reality and rationality. In doing so, it has made significant strides toward turning science into spin and sapping the power of knowledge to ameliorate the human condition.

CHAPTER 10

The Dismal Science Is Certainly Dismal, but Is It Science?

ECONOMICS HAS A PARTICULARLY DISMAL RECORD as a science in the United States since the dawn of the Cold War, but that's not where its reputation as "the dismal science" comes from. That owes to the perception of its earliest practitioners as relentless purveyors of Malthusian gloom.

Economic science is the field of knowledge concerned with how goods and services are produced and distributed to serve the needs of human populations. The wealth and poverty of nations and individuals are what it seeks to understand, with the presumed goal of improving the material condition of the human race. Despite the misuse economics has suffered in American universities and policy circles, the answer remains: Yes, it is a science. If practiced as it should be, it could arguably be the science of greatest benefit to humankind. At present, however, it serves to tie all the strings of the corporate dominance of science together.

The Crucial Role of Criticism in Economic Science

Social sciences, including economics, differ from the natural and physical sciences in a fundamental way. Their practitioners can't place themselves outside and apart from human society, which is what they are studying. Intimately entangled in their subject matter, how can social scientists stake a claim to scientific objectivity?

Philosopher of science Sandra Harding has argued that objectivity in the social sciences requires that they be "critical and self-reflective."[1] To be self-reflective, a science's practitioners have to continuously monitor their own practices to keep financial incentives and ideological biases from influencing their scientific judgment. And that can't be done individually; it has to be a *social* endeavor encompassing the collective, collaborative efforts of large communities of researchers.

As for the critical element, scientific objectivity is unthinkable in the social sciences without free and open debate among contending schools of thought. The history of the economics profession over the past century reveals a grave deficiency in that regard, as legitimate theoretical perspectives have been sharply restricted or suppressed in their entirety. But it hadn't always been that way.

Eighteenth Century: Laissez Faire!

Economic knowledge took great strides forward in the eighteenth century, culminating in a new paradigm presented by a Scotsman, Adam Smith, in an enormously influential book entitled *The Wealth of Nations*. Smith happened to be living in the era of earthshaking social transformation that came to be known as the Industrial Revolution. By analyzing the revolutionary processes he was witnessing, he was able to offer a coherent, satisfying theory of how society could best organize its productive forces to maximize social well-being.

Smith had drawn inspiration from a school of French social philosophers, the Physiocrats, whose teachings were commonly reduced to the familiar slogan *Laissez faire!* Let it be. Allow producers and consumers, sellers and buyers, to freely exchange goods without interference from the ruling powers. Although Smith himself did not use the famous phrase, posterity has adopted it as shorthand for free-enterprise ideology.

The explanatory success of Smith's paradigm gave it an aura of universal truth on a par with Newton's discovery of the law of universal gravitation. With the passage of time, however, and the further development of British industry, it became evident that free-market prescriptions were not eternally and universally applicable in a rapidly changing world. Smith's theories were fair game for criticism.

Nineteenth Century: Criticism Intensifies and Economic Science Flourishes

Discussion and debate over the ideas of Adam Smith and his successors stimulated the growth of economic science throughout the nineteenth century. The most systematic and consequential critique was offered by Karl Marx, who admired Smith as a pioneer of "classical political economy." The adjective "classical," however, placed it in historical context and challenged any claims on its behalf to timelessness. As brilliantly as Smith had analyzed the economy of 1776, the world he described had radically changed and would continue to change.

Karl Marx's critical analysis of the capitalist system attracted fervent supporters and opponents alike. But as long as the debate remained an open one, Marx's adversaries could not simply ignore his critique; they were obliged to engage with his ideas.

Twentieth Century, Part I: The Debate Transcends the Purely Theoretical

Early in the twentieth century, the intensifying rivalry of imperialist nations engulfed the world in war. By war's end, one major combatant, Russia, had exited the capitalist system to embark on a novel postcapitalist path wherein supply and demand no longer reigned supreme. The Russian Revolution had presented economic science with a global experiment testing two incompatible hypotheses regarding the organization of economic affairs nationally and internationally.

Economic policy in the newborn Soviet Union was inspired by Karl Marx's theories. Ten years into the great experiment, however, the Old Bolsheviks who had led the Russian Revolution were gone—physically exterminated by Stalin's faction. Official Marxism subsequently devolved into empty slogans justifying Stalin's shifting domestic and foreign policies.

In the United States, meanwhile, the fearsome specter of Bolshevism had made any brand of anticapitalist economic theory anathema to the political establishment. Proponents of Marxist ideas were discriminated against and excluded from university economics departments and other institutions of the economics profession. Serious consideration of alternatives to capitalism was ruled out of bounds.

The upshot was that economic science, both East and West, ceased to be animated by the spirit of free and open debate. The door to a critical and self-reflective social science had slammed shut.

The Great Depression Shook Things Up

The practice of economics as a science in the United States would not hit rock bottom until the Cold War, but its steep decline commenced decades earlier. With the American capitalist economy riding high in the Roaring Twenties, academic economists paid little attention to critiques of laissez-faire economics. The euphoria, however, was gone by the end of the decade.

The Great Depression of the 1930s rocked the capitalist economic system to its foundations. Before that cataclysmic event, the predominant view among professional economists in the West was that capitalism's ups and downs were eminently manageable and certainly nothing to worry about. The Depression forced a radical rethinking of that premise. Marx's critique of capitalism gained a renewed hearing, but the confusion caused by its association with Stalin's policies in the Soviet Union put it at an insurmountable disadvantage.

The Advent of Keynesianism

Meanwhile, British economist John Maynard Keynes offered a new explanation of what had gone wrong and what to do about it. His ideas provided the basis for a school of economic thought that came to bear his name: "Keynesianism." Keynes and his followers placed the blame for the economic meltdown of the Great Depression on *inadequate aggregate demand*. When manufacturers cannot sell their products because too few consumers have the wherewithal to buy them, production is curtailed and the economy goes into a tailspin. The Keynesian solution to this problem is government spending to create demand that keeps the wheels of the economy turning.

Keynes was not merely advocating old-fashioned "prime the pump" policies wherein limited amounts of temporary government spending would jump-start a sluggish economy. In his view, the amount of government spending necessary to keep a modern capitalist economy on the tracks would have to be *massive* and *never-ending*. Taxation would not be sufficient to sustain the level of government funding required by this strategy. The extra amount had to be provided by *deficit spending*. And to be successful, the deficit spending would have to not only continue without end but also *continually increase*.

"But Mr. Keynes," he was asked, "how is that possible? How can a national economy endure larger and larger budget deficits forever? *What will happen in the long run?*"

"In the long run," Keynes replied, "we are all dead." That was his way of saying the problem was for future generations and their economists to deal with.[2]

Keynes won the ear of Franklin D. Roosevelt, who as president of the United States had the authority to put his ideas into practice. Roosevelt's New Deal policies were Keynesian in spirit, but his government's deficit spending in the 1930s was far too timid to lift the American economy out of depression. What finally pulled it out of the mire was the enormous volume of government deficit spending forced upon Roosevelt by World War II. War spending produced immense aggregate demand, millions of unemployed workers were put back to work, and the Depression rapidly ended.

After the war, Keynesians were in the majority in university economics departments. A stubborn minority, however, disliking Keynes's insistence that the private sector must ultimately rely on the public sector for survival, damned him as a "socialist." The epithet was unjustified. Keynes was in fact devoted above all else to saving the capitalist system from destroying itself. With the onset of the Cold War and the rise of McCarthyism, however, the red-baiting tactic contributed to weakening the Keynesians' influence.

Twentieth Century, Part II: Cold War Puts Economic Science in Deep Freeze

The rigor mortis afflicting economic science reached total paralysis after World War II. Academic economists could not objectively discuss non-mainstream ideas without their being accused of harboring Communist or pro-Soviet sympathies and facing removal from their university posts. The majority succumbed to enforced conformity and self-censorship, applying their considerable intellectual talents not to disinterested pursuit of objective knowledge but to accepting without question the eternal superiority of the international market-based economic system.

Their research agendas were confined to problems *within* the free-enterprise paradigm and eschewed any new knowledge that might contradict it. It thereby complemented the subordination of other sciences to corporate profitability with a theoretical framework of apologetics for the profit system as a whole.

Keynesian Sunset and Chicago School Dawn

As military spending declined following the war, a high level of aggregate demand was sustained by the Marshall Plan's enormous expenditures in the rebuilding of Europe. With the passage of time, as prosperity returned, so did complacency.

By the early 1950s, memories of the Depression years had faded, and private investors increasingly saw federal spending not as their savior but only as unwanted competition. Their political allies turned against New Deal policies, and eventually partisans of extreme promarket ideology displaced the Keynesians in the halls of academia.

The anti-Keynesians, led by Milton Friedman and Friedrich Hayek, were called *neoclassical* economists, but a more appropriate label, in today's context, might be *protolibertarians*. As early as 1944, Hayek put forward a radical free-marketeering position in the polemical book he is best known for, *The Road to Serfdom*. Because Friedman and Hayek were associated with the University of Chicago, they and their followers were dubbed the Chicago School of economists.[3] Keynesian ideas retained some clout into the early 1970s, but from that time forward the rise of the Chicago School to dominance in university economics departments was unrelenting.

Ideological Polarization and the Vietnam War

In the 1960s, an ideological divide in the United States opened and continued to widen. A crucial accelerant was a regional conflict in Southeast Asia that rapidly moved to center stage in world affairs.[4] Supporters of US involvement in Indochina were furious that millions of young Americans were marching in the streets, protesting their own country's military aggression.

The antiwar youth radicalization created ferment in the universities that led to a weakening of the Cold War conformity within economics departments. Criticism of the dominant neoclassical paradigm increased, even to the point of allowing a few Marxist economists to join the debate.

At the same time, however, defenders of the free-market faith doubled down. In his 1964 presidential campaign, Barry Goldwater famously extolled the virtues of immoderation—"Extremism in defense of liberty is no vice"—and the conservative movement has become more and more extremist—which is to say, *less conservative*—ever since.

Prowar intellectuals, derided as "Cold War liberals" or "State Department liberals," began to question whether they were liberals at all, and many concluded they were not. The upshot was a momentous expansion of the conservative intelligentsia. These defectors from liberalism, led by public intellectuals including Irving Kristol and Norman Podhoretz, dubbed themselves *neoconservatives,* and a new ideological movement was born. The neocon contribution to the dismal science was spirited sponsorship of the doctrine of supply-side economics—the unsupported assertion that drastically reducing taxes on corporations and the super-rich will stimulate economic growth to the benefit of the entire American population. Irving Kristol, known as the Godfather of Neoconservatism, glibly admitted he "was not certain" of supply-side theory's "economic merits but quickly saw its political possibilities."[5]

And then along came an even more reactionary and intransigent cult.

Not Your Grandparents' Libertarians

Once upon a time, not too many years ago, there was more diversity in libertarianism. Like most ideologies, it comprised a variety of positions and principles. Some libertarians voiced concerns traditionally associated with progressive causes, such as privacy rights, civil liberties, and antimilitarism.

But in the 1980s, with the encouragement and material support of Margaret Thatcher's government in Britain and Ronald Reagan's in the United States, a virulent strain of libertarianism became an ever-growing presence. The new hardcore libertarians have marginalized their gentler predecessors, deriding them as impotent "conventional libertarians." When today's libertarian movement trumpets its devotion to individual rights, it is code for *individual property rights* and has nothing to do with the *human rights* of the vast majority of individuals. In this ultralibertarian worldview, an individual without property has no rights.[6]

The triumph of the new libertarians was by no means a spontaneous development or a case of superior ideas driving out erroneous ones. To understand how it came about, follow the money. Right-wing scholars devised the theory and tactics, but it was the Koch brothers' coalition of some five hundred ultrawealthy donors that "made the Word flesh" and turned a small, fanatical cult into a megachurch.

Goodbye GOP, Hello KDN

The traditional Republican Party has been replaced in all but name with an ultra-right-wing party that in truth should call itself the KDN: the Koch Donors Network.[7] In 2017 this stealth party utilized Donald Trump as its accomplice in passing tax and budget legislation that transferred $1.5 trillion from average American families into the bank accounts of the wealthy.

In the political sphere, the KDN "rivals the Republican National Committee in size, scope, and budget." As of 2017, its "grassroots organizing arm," Americans for Prosperity, had chapters in dozens of states.[8]

The Koch coalition holds annual conclaves where several hundred of America's wealthiest gather to receive obeisance from leading senators, congresspeople, and governors. At the 2017 KDN meeting, a spokesperson announced that the group planned to spend from $300 million to $400 million to influence the 2018 elections. The day before that meeting, Charles Koch met privately with vice president Mike Pence, the donor network's principal liaison with the Trump administration.[9]

Although the provenance of the individuals chosen to staff Trump's White House offices and agencies at all levels cannot always be traced, they are not a random assortment of right-wing political careerists. A significant proportion are cadre trained by the libertarian think tanks funded by the Koch network. That was accomplished by the large number of Heritage Foundation operatives, including the think tank's longtime leader Edwin Feulner, who did the choosing. An investigative journalist described how Trump's "transition team" functioned: "Heritage employees have been soliciting, stockpiling and vetting resumes for months with an eye on stacking Trump's administration with conservative appointees across the government."[10]

The "Small Government" Swindle

The well-funded libertarian research institutes are single-mindedly devoted to "dismantling the administrative state." As right-wing antitax crusader Grover Norquist famously exclaimed, "I don't want to abolish government. I simply want to reduce it to the size where I can drag it into the bathroom and drown it in the bathtub."[11]

While denouncing "statism" and all governmental influence on the economy, they allow one enormous exception: they depend on the power of the state to defend the property rights upon which their notion of liberty is based. The contradiction is profound: while demanding total destruction of governmental power, they nurture the most powerful military state—or "national security state"—the world has ever seen.

The Conspiratorial James McGill Buchanan

The libertarians' ruinous impact on economics is best illustrated by the seminal contributions of James McGill Buchanan. Buchanan was awarded the Nobel Memorial Prize in Economic Sciences in 1986, but his practice of the discipline made a mockery of the very notion of economics as a science.

As one of the Chicago Scholars who had taken the laissez-faire express to extreme libertarianism, Buchanan had arguably become the ideology's most effective advocate. Despite his stature as a Nobel laureate, though, his name is not well known to the general public. Why?

Buchanan was not a publicity hound. He didn't broadcast his views far and wide, because he never wanted them to be widely known. He believed that certain vital truths about the world we live in should be hidden from public view. After his death in 2013, his private papers revealed warnings to co-thinkers that "conspiratorial secrecy is at all times essential."[12]

Buchanan's secret truth was that democracy and "liberty" are incompatible, and that therefore democracy must be suppressed. "Despotism," he once wrote, "may be the only organizational alternative to the political structure that we observe," by which he meant the system defined by the American Constitution.[13]

Makers versus Takers

In Buchanan's worldview, the population is divided (to use Ayn Rand's colorful terminology) into *makers* and *takers*. The makers are the productive classes—owners of capital whose profit-making activities expand the national economy—and the takers are the indolent masses. To Buchanan, any taxation that redistributes wealth from the makers to the takers is a downright immoral form of robbery, and any governmental attempt to regulate the makers' businesses is a criminal violation of their liberty.[14]

The economic history of the world is indeed a story of takers robbing makers, but Buchanan's interpretation has the relationship upside down and backward. The United States' great prosperity derived first of all from the agricultural production of African slaves—Abraham Lincoln called it "the wealth piled by the bondsman's 250 years of unrequited toil"—and secondarily from the industrial production of the underpaid labor of industrial workers. A small number of southern plantation owners and northern manufacturers amassed vast and enduring family fortunes by appropriating the profits produced by those laborers. Who, then, were really the makers and who were the takers?

The ill-gotten wealth of the exploiters of labor allowed them to gain political control, limit the franchise of the laborers, and create a legal system to consolidate their system of economic injustice. Adding insult to injury, the slaveholders and robber barons justified their conquest by propagating ideologies—from Social Darwinism to libertarianism—that denied and devalued the laborers' role in creating the modern American economy.

To appreciate the audacity of Buchanan's perversion of history, consider the plight of the former slaves after the US Civil War. Having been forcibly *taken* from their homelands, having had their labor violently *taken* from them for decades, and being left in dire poverty in the postbellum South, many were dependent on bare-bones federal assistance for survival. That made them, in Buchanan's eyes, contemptible "takers."

The Lenin of Libertarianism?

What places Buchanan among the most dangerous of the self-styled economic scientists is that he not only professed antidemocratic ideas but also devised strategies to successfully implement them. He was a social engineer who found ways to turn right-wing economic theory into public policy. It has been suggested that as the movement's key cadre-builder, Buchanan was to libertarianism what Lenin was to Marxist socialism.[15]

Buchanan took the ideas he learned from his Chicago School mentors to the University of Virginia and created a more extreme Virginia School of economics. Its institutional expression was the Thomas Jefferson Center for Studies in Political Economy, which he founded in 1957 to develop "a line of new thinkers" to challenge the "increasing role of government in economic and social life."[16] That was to be accomplished by a "constitutional revolution" that would covertly rewrite the rules of the American economy to enrich the few at the expense of the many. Among its primary ambitions were the total elimination of the social security, public health, and public education systems.

Buchanan's research center was the embryo of the modern libertarian movement. He envisioned the creation of a "counterintelligentsia" backed by a "vast network of political power" to replace the existing establishment intellectuals.[17] He thus provided the blueprint for today's powerful array of libertarian think tanks and their army of paid academics and lobbyists.

Buchanan's plans would have languished on the drawing board without the material support necessary to put them into action. In 1983 he reconstituted his academic institute at George Mason University, renaming it the Center for Study of Public Choice. George Mason University, identified in the *Wall Street Journal* as "the Pentagon of conservative academia,"[18] was the ideal venue for Buchanan's operation.

As mentioned, the Koch brothers have donated tens of millions of dollars to George Mason University and to Buchanan's Center for the Study of Public Choice, which trained the young intellectuals who would fill the Koch think tanks, become speechwriters for Koch-financed congressmen, and staff the Trump administration. Eventually, tactical disagreements led the impatient billionaire brothers to force Buchanan out (he "retired" in 2007) and take direct control of the research center.

An Assessment of Buchanan's Scientific Praxis

The science promoted by Buchanan's economics research centers is founded not on objective premises but on the moral judgment that the vast majority of human beings are economic parasites on the capitalist class. Furthermore, Buchanan's public choice theory—the work for which he was awarded the Nobel—is fundamentally unscientific in its reductionist methodology.

Public choice theory reduces real-world economic decision-making to the sterile abstractions of mathematical game theory. In a universe where human beings always act like purely self-interested automatons, game theory could perhaps offer some useful insights into economic behavior.[19] But Buchanan applied mathematical models based on misanthropic assumptions about human nature to complex social interactions.

Historian Nancy MacLean has described the hypothetical social order from which public choice theorists deduced their axioms as one in which "individuals always acted to advance their personal economic self-interest rather than collective goals for the common good." Buchanan and his fellow theorists, she says, were simply conducting

> thought experiments, or hypothetical scenarios with no true research—no facts—to support them, while the very terms of their analysis denied such motives as compassion, fairness, solidarity, generosity, justice, and sustainability.[20]

In brief, Buchanan's method is designed not to attain new knowledge about economics but to justify an economic system of vast material inequality. It is not science but spin.

Can Economic Science in America Be Raised from the Dead?

At the outset of this chapter, I stated that the presumed goal of economic science was to improve the material condition of the human race. That is clearly not the purpose of the economics profession in the United States today, as its idols of "growth" and "austerity" drive an ever more unjust and unsustainable maldistribution of wealth.

Restoring American economic science to health would first of all require a social environment conducive to the kind of free and open debate that scientific objectivity requires. Safeguarding the all-important self-critical element of the science also requires strong institutional norms, public oversight, and transparency. Only when a reoriented economics achieves all that will it deserve to be called a science.

The
Militarization
of
American
Science

CHAPTER 11

Science Harnessed
to the Chariot of Destruction

WHAT IS SCIENCE "FOR"? Some would say it is for increasing humanity's storehouse of knowledge, with no qualifications imposed on the utility of that knowledge. But most of us also expect science to generate knowledge that will in myriad ways prove beneficial to the human race. We hope, at the very least, that it will not be detrimental to human health, happiness, and general well-being.

Unfortunately, the trillions of dollars the American government spends in the name of science is primarily devoted to perfecting military technologies of death: finding new and better ways to kill people in large numbers.[1] Before dismissing this claim as hyperbole, consider the evidence:

- "Trillions of dollars": From the end of the Vietnam War through 2017, according to official sources, government science expenditures totaled more than $2,859,000,000,000—almost $3 trillion.[2]

- "Primarily devoted": "More than half the U.S. government's spending on R&D is for national defense."[3]

As for the ultimate purpose of these expenditures, the military's raison d'être is to wage war, and the focus of war is the killing of enemy combatants. "I'm not subtle," Trump's former secretary of defense James "Mad Dog" Mattis declared, "I need to make the military more lethal."[4]

Those trillions of dollars have created a growing stockpile of nuclear and high-tech weaponry. Lobbyists for the war industry and the

140

politicians who serve them would have us whistle past that vast grave-yard, ignoring or denying the criminality and the danger.

The Dimensions of American War Spending

A "staggering number of tax dollars," states a specialist in the economics of military spending,

> is subsidizing the US military-industrial complex, year in, year out, at levels that should be (but aren't) beyond belief. In 2019, Pentagon spending is actually higher than it was at the peak of either the Korean or Vietnam conflicts and may soon be—adjusted for inflation—twice the Cold War average.[5]

In July 2019 the Republican-controlled Senate and the Democrats in the House of Representatives agreed that the 2020 federal budget would include $738 billion for the military.[6] That would represent about half of all federal discretionary spending. Those figures, however, do not include *all* military spending. The cost of past wars—the significant fraction of interest on the national debt that military deficit spending has generated, plus what the Department of Veterans Affairs spends—should also be added. So should the budgets of the State Department, the Homeland Security Department, NASA, and a few other agencies. With all that, the military share rises from half to two-thirds, and the absolute amount spent on war doubles to about $1.25 trillion a year.[7]

Divide $1.25 trillion by 365 and you'll find that the United States is spending more than $3.4 billion a day on past, present, and future wars. That's $140 million an hour, or almost $40,000 a second. Pause a moment to let that sink in. And average Americans wonder why we pay so much in taxes and get so much less of value in return!

NASA's budget should be included in war spending because the agency's costly space exploration activities and aeronautical research represent military spending in disguise. NASA's endeavors are funded not to make space enthusiasts happy but to satisfy the needs of the Pentagon and the industries that supply it. American military strategists treat "outer space" as a crucial arena of international competition, and they aim to control it so China and other potential rivals can't. Any remaining doubts on this score should have dissipated with Trump's announcement that a space force would become a separate branch of the armed forces.[8] And although nuclear weapons are generally counted as military expenses, so-called Atoms for Peace programs are not. They should be.[9] American politicians' vociferous protests against Iran's domestic nuclear program amply expose the military implications of power plants generating electricity by nuclear fission.

Is it unreasonable to count the State Department's budget as war spending? Doesn't diplomacy, the State Department's stock-in-trade, focus on *avoiding* war? Consider this: the diplomatic efforts of the State Department include forming military alliances such as NATO, creating "coalitions of the willing" to attack other countries, and organizing economic blockades that also kill people. When secretary of state Madeleine Albright was asked about a United Nations report that 567,000 Iraqi children under the age of five had died as a result of the economic sanctions she oversaw, she replied, "We think the price is worth it."[10]

Furthermore, as policy critic Chalmers Johnson observed, the American embassies in many countries are not "true embassies" but "walled compounds, akin to medieval fortresses, where American spies, soldiers, intelligence officials and diplomats try to keep an eye on hostile populations in a region at war."[11] Finally, no attempt to calculate how much money flows to the American military can ignore the shocking extent to which Pentagon spending is unaccounted for.

At Least $6.5 Trillion in Pentagon Spending Is Untraceable

Although legally bound by the same accountability requirements that all federal departments are required to meet, the Department of Defense has simply ignored the law for decades, and has gotten away with it. That is not an unfounded allegation of antimilitarist critics; it is the official conclusion of the DoD's own Office of the Inspector General, which stated in a 2016 report that the Pentagon's financial statements "were unreliable and lacked an adequate audit trail."[12] As a result, over a number of years the US Army had spent *six and a half trillion dollars* that had not been properly accounted for.[13] And that $6,500,000,000,000 represented only the *army's* untraceable funds. How many more trillions were likewise not accounted for by the navy, air force, and marine corps?

The Pentagon's lack of accountability is not simply a matter of sloppy bookkeeping. It serves to hide slush funds for activities the American public has not approved of and would not approve of—black sites, black ops, secret wars, and "ghost militias" operating in some 135 countries throughout the world.[14] With regard to science spending, that would include funding for programs like BSCT (behavioral science consultation teams), designed to optimize the use of torture in interrogations at Guantánamo Bay and elsewhere.[15]

What Was It All For?

In 1789 the United States government established a Department of War. The name endured for the better part of two centuries. Two years after World War II, however, the agency dropped "War" from its name and in 1949 officially became the Department of Defense.

The rebranding succeeded in its purpose. In the paranoiac Cold War context, the American public was conditioned to accept the limitless military buildup as a defensive necessity. Meanwhile, the aggressive campaign of McCarthyism assured minimal dissent. The defensive posture was fraudulent. Since World War II, the American military machine has been deployed throughout the world not to defend against attacks on the United States or its people but to aggressively extend the economic dominance of corporate America. Britannia had ceased to rule the waves; the mantle of imperial power had passed to the United States.

It was imperialism of a different stripe, however; a difference acknowledged in the need for new terms, like *neocolonialism* and *neoimperialism*, to describe it. Unlike the older forms based on direct territorial conquest, the updated style would exercise control more efficiently by indirect financial and economic means, enforced by the constant threat of military violence looming in the background.

The previous chapter described how one social science, economics, had been enlisted to justify the growing corporate stranglehold on American society. Providing theoretical justification for the postwar world order would require the complicity of another branch of the social sciences.

Enter the Political Scientists and the American Century

The social science nicknamed poli-sci on American campuses was instrumental in devising and promoting an ideology to bolster the new imperial ambitions. In the early postwar years, the expression of those ambitions took public form as a quest for "the American Century." Later, neocon political scientists and their think tanks would give it their own special twist.

The United States emerged from World War II as the only significant power with its industrial base intact. Its enemies' economies lay in ruins, as did those of its allies. American corporations appeared to have a clear path toward unchallenged rule over the world capitalist market, backed by military domination based on a monopoly of nuclear weapons. The American Century seemed to be at hand.

There was, however, one major obstacle to American global hegemony. A significant portion of the international economy was *not* capitalist—the part controlled by the Soviet Union. The Soviet state, however, had been pounded so hard by Hitler's armies that it was deemed incapable of withstanding the American juggernaut. The number one item on the American Century's wish list was reopening the Soviet economy to capitalist investment. That was the motive force driving the Cold War.

The Soviet Surprise

To the chagrin of American policymakers, the USSR's planned economy proved far more resilient than free-market ideologues had thought possible. Rather than collapsing, it rapidly rebuilt its heavy industry and became a rival superpower to the United States.

The Soviet Union likewise rose to the rank of superpower in the world of science, allowing it to break the US nuclear—and then thermonuclear—monopoly. Its scientific prowess was further demonstrated by launching the first human-made satellite into orbit and then sending the first human, Yuri Gagarin, into space.

Far from shrinking and disappearing, the portion of the world closed off from American investment began to expand. Within three years of the war's end, half of Germany and all of Poland, Czechoslovakia, Hungary, Romania, Yugoslavia, Bulgaria, and Albania had become integrated into the Soviet economic bloc. Then, one year later, in 1949, came the biggest blow of all: the Chinese Revolution removed a quarter of the world's population from the realm of the free market. American policymakers bemoaned "the loss of China."

The Chinese example inspired the rise of national liberation struggles against American neocolonialism throughout Asia, Africa, and Latin America. In 1959 an island just ninety miles from America's shores defiantly exited the world capitalist order. The dream of a global *Pax Americana* ruled by an all-powerful market-based empire was dead . . . but by no means forgotten.

In the early 1990s, the tide turned, dramatically. The dissolution of the Soviet Union and related events in East European countries put an end to centralized governmental control of their economies, thus reopening them to private investment and the international capitalist market. China, while retaining centralized political control, began to allow the growth of a substantial private sector and large-scale foreign investment.

The New American Century

The Cold War had ended with the United States as the last superpower standing. The dream of a global American empire revived, and neocon political scientists were in the vanguard of striving to make it a reality. William Kristol, scion of the first family of neoconservatism, was among the ideological leaders of the charge. He and fellow neocon Robert Kagan laid out their vision in a seminal article in *Foreign Affairs*, calling for the United States to establish a "benevolent global hegemony." Fully aware that the benevolence would have to be imposed by force, they recommended raising military spending by $60 to $80 billion a year.[16]

In early 1997, Kristol gave his program organizational heft by founding the Project for the New American Century (PNAC). He enlisted such luminaries as Dick Cheney, Donald Rumsfeld, Paul Wolfowitz, and Jeb Bush[17] to sign PNAC's Statement of Principles, which declared, "We need to increase defense spending significantly if we are to carry out our global responsibilities today and modernize our Armed Forces for the future."[18] Most ominous was the manifesto's call to "shape circumstances before crises emerge, and to meet threats before they become dire," an unambiguous formula for preemptive military action.[19] The New American Century required leaders who would not be squeamish about unilaterally exercising military power.

Desire for a militarized New American Century was bipartisan. That was demonstrated in deeds as well as words. From the end of the Cold War through 1999, with Bill Clinton as commander in chief, "the United States has embarked on nearly four dozen military interventions . . . as opposed to only 16 during the entire period of the Cold War."[20] Clinton's secretary of state, Madeleine Albright, captured the spirit of the New American Century by demanding, "What's the point of having this superb military if we can't use it?"[21]

Think tanks across the American political spectrum clambered aboard the bandwagon. The bipartisan Council on Foreign Relations had been the first to showcase Kristol's neoimperialist program in its journal, *Foreign Affairs*, and the Democratic Party's Progressive Policy Institute (PPI) echoed PNAC's advocacy of American global domination.[22]

In January 1998, PNAC had issued an open letter to President Clinton, demanding that he "undertake military action" to remove Saddam Hussein's regime from power to preserve the stability of "the entire middle east."[23] The events of September 11, 2001, gave PNAC the rationale it needed to win public opinion to its campaign for a "muscular American foreign policy." The destruction of the World Trade Center towers by a minuscule group of Islamicist fanatics led to widespread calls for revenge and a modern crusade against the Muslim world.

Think tanks ranging from Brookings to the American Enterprise Institute conducted an unrelenting drive to persuade policymakers to prosecute wars and the American population to support them. PPI joined hands with PNAC to bring about a national consensus of the political elites in favor of military action. Defeating the "enemies of modernity," they insisted, would require the United States to impose its leadership on the entire world, using military force as necessary.

The Trump Bumps on Military Spending (Upward) and Science (Downward)

Donald Trump campaigned for the presidency on a promise to sharply increase military spending. That was a promise he kept. For starters, he added $30 billion for the War on Terror to Obama's proposed 2017 budget and then another $127 billion to his own administration's 2018 budget.[24]

As a *New York Times* editorial lamented, while Trump proposed draconian spending cuts to health care, Medicare, Social Security, and other social programs, "the Pentagon remains a sacred cow, destined to receive yet more money."[25] The official 2016 military budget of $584 billion went up to $643 billion in 2017. But Trump's initial proposal for $677 billion for 2018 was not enough to satisfy military-besotted legislators, Democrats and Republicans alike. "In a rare act of bipartisanship on Capitol Hill," the Senate voted 89 to 9 to approve a $700 billion Military Authorization Bill "that far exceeds what President Trump has asked for."[26] Trump signed the bill into law on December 12, 2017.

Trump's well-documented disdain for science makes it unsurprising that while the military's share of federal research money would rise, nonmilitary science and technology spending would decrease significantly. "Under Trump's spending blueprint, the Pentagon's R&D budget for fiscal 2018 would be $84.9 billion, including war spending, compared to $76.4 billion in fiscal 2017." Meanwhile, research spending by nondefense departments and agencies would decline, and "in some cases it would dip precipitously. For example, the EPA's R&D budget in fiscal 2018 would be fully 46 percent lower than in fiscal 2016, while the Agriculture Department's spending on such programs would be 25 percent lower for the same period."[27] The budget of the National Institutes of Health was likewise slashed by 22 percent.

Weaponized Keynesianism

What can account for the inexorable growth of American war spending? One explanation that deserves serious consideration is "weaponized Keynesianism," a term coined by a progressive congressman, Barney Frank:

> We have a very odd economic philosophy in Washington. It's called weaponized Keynesianism. It is the view that the government does not create jobs when it funds the building of bridges or important research or retrains workers, but when it builds airplanes that are never going to be used in combat, that is of course economic salvation.[28]

Barney Frank's synopsis of the Keynesian position is accurate, but by calling it an "odd economic philosophy," he suggests that it is a false, subjective idea rather than an objective economic reality. Frank believes out-of-control military spending is merely an erroneous policy choice, which would make it a solvable political problem and not a permanent feature of the current American economic system. His proposed solution is to curb the political influence of the Pentagon and weapons contractors. In other words, he doesn't really think Keynesianism has anything to do with it; he uses the term ironically.

The collusion between generals, merchants of death, and politicians that Frank decries is an obvious driver of ever-growing war budgets, but is that the prime mover or, as Keynesian theory suggests, a superficial expression of an underlying reality? A straightforward, as opposed to ironic, interpretation of weaponized Keynesianism maintains that the economic system itself depends upon an astronomical volume of war spending for survival.

That contention is worth examining. Because the form, content, and direction of science have been so strongly influenced by expenditures on war-related research, knowing why that money is spent is essential to understanding the place of science in contemporary America.

Revisiting Keynes

Keynesianism is a theory that purports to explain why economic crashes like the Great Depression of the 1930s occurred.[29] It is based on a hypothetical construct, which is to say, an unproven postulate that must be assumed true if the overall theory is true. In the case of Keynesianism, the hypothetical construct is that economic crises are caused by "insufficient aggregate demand." As a corollary of that theory, weaponized Keynesianism depends on the same hypothetical construct.

Whether or not the American economy is incurably addicted to war spending cannot be proven by deduction from Keynesian axioms. But weaponized Keynesianism offers a highly plausible explanation for observable macroeconomic phenomena that can't simply be ignored. The Great Depression demonstrated that the invisible hand of supply and demand could not be counted on to restore an immobilized economy to equilibrium. Keynes's insights led to economic reforms, including such anticyclical devices as Social Security and unemployment insurance to soften the economic effects of diminishing purchasing power when

workers lose their jobs. But the crucial plank in the platform of Keynesianism was the need for massive federal deficit spending to sustain aggregate demand.

Roosevelt's public works programs, however, entailed far too little
deficit spending to revitalize the American economy. What really did
the trick was unprecedented, massive military expenditures. As economist Paul Krugman observes, it was "the large public works program,
otherwise known as World War II, that ended the Great Depression."[30]
After the war, to prevent the world economy from once again lapsing
into a terminal crisis of overproduction, deficit spending would have to
not only continue without end but also continually increase. The history
of ephemeral "peace dividends"—failed attempts to introduce a note of
rationality into the irrational addiction to arms expenditures—serves to
illustrate the tenacity of weaponized Keynesianism.

The Post–World War II Peace Dividend
The defeat of Germany and Japan in 1945 was greeted with jubilation by
most Americans but not quite all. The barons of the weapons industry
who had grown fat on military contracts saw themselves facing a sudden
starvation diet. The war-weary American public was in no mood to continue bankrolling martial glory. That would change, however, with the
advent of a deliberate campaign "to scare the hell out of the American
people." That oft-quoted line from senator Arthur Vandenberg is said to
have prompted the proclamation of the Truman Doctrine, which promised military support to all "free people who are resisting attempted sub-
jugation by armed minorities or by outside pressures."[31]

The enemies of free people, it turned out, were limited to a specific
variety: those associated with the Soviet Union. The Truman Doctrine
thus gave birth to the obsessional Cold War anticommunism that would
serve for more than four decades to justify weaponized Keynesianism.

The Post–Vietnam War Peace Dividend
The anticommunist crusade peaked in the McCarthy years and then
lost much of its potency due to the youth radicalization sparked by the
Vietnam War. When that conflict ended, some progressive politicians

proclaimed that ever-increasing war budgets were no longer necessary. Another peace dividend was in the offing.

And indeed, after 1973, military expenditures began to decline; by 1980 they had decreased by as much as 25 percent. But the addicted economy was not to be denied its fix for long. Ronald Reagan's election in 1980 ushered in a swift rebound. By the end of his first term, military spending had surpassed the levels reached during the Vietnam War and continued climbing. Anticommunism as a fear factor made a comeback.

The Post–Cold War Peace Dividend

But toward the end of the 1980s the Cold War began to thaw, and it seemed to have completely disappeared with the dramatic fall of the Berlin Wall in November 1991. Massive funding of science and technology would no longer be required to keep pace in an arms race with a defunct Soviet Union. Yet another peace dividend was widely anticipated, and the Pentagon's budget did briefly decline by a few percentage points. But once again, policymakers could not allow this situation to persist. A new rationale for high levels of R&D spending was needed, but unrestrained military budgets remained a hard sell for legislators until September 11, 2001. The War on Terror provided permanent war as the perfect justification for a permanent war economy.

The bottom line is that the free-market, profits-driven economy recognizes only one sufficient anticyclical device: the enormous artificial market for high-tech research and industrial production created by hundreds of billions of dollars a year in military spending. As would be expected, this is an extremely sore point for mainstream apologists, who vehemently deny it. Their insistence that the Pentagon budget has a negligible impact on employment is an exercise in obfuscation and the misuse of statistics. In fact, the weapons industries directly and indirectly sustain an estimated 13 percent of the manufacturing employment base of the United States.[32]

Add in the "multiplier effect" that creates millions of service jobs, all the way down to the cashier at the diner next to the missiles plant or the army base. And don't forget the more than two million active-duty and reserve servicemembers (none of whom are counted in official employment statistics, by the way). When all of this is taken into account, the extent to which American jobs, incomes, and economic demand depend on a permanent war economy is hard to miss.

Jobs-Junkie Geography

The evidence of addiction to jobs creation by military spending is all over the map—both the US map and the world map. In the United States, Pentagon money is deliberately spread to as many congressional districts as possible in order to assure voter support for its multibillion-dollar budgets. Investigative journalist James Fallows offers this typical example:

> In the late 1980s, a coalition of so-called cheap hawks in Congress tried to cut funding for the B-2 bomber. They got nowhere after it became clear that work for the project was being carried out in 46 states and no fewer than 383 congressional districts (of 435 total).[33]

As for the world map, in 2015 the Pentagon had, by one account, "some 562,000 facilities worldwide, which collectively take up 24.7 million acres, or nearly the size of Virginia."[34] This was the American "Empire of Bases" so thoroughly described and analyzed by Chalmers Johnson.[35]

American weaponized Keynesianism is also displayed on the map of international weapons sales. The United States has long been by far the world's largest arms dealer, dominating the markets for warplanes, bombs, and missiles. Weapons producers like Lockheed Martin, Boeing, Raytheon, and General Dynamics are primarily sustained by the Pentagon cash cow, but the American employment rate gets an extra boost when the merchants of death peddle tens of billions of dollars' worth of their wares to Israel, Saudi Arabia, Egypt, Turkey, and any number of other countries.

Why Deficit Spending on Useful Things Can't Stop Depressions

As Barney Frank indicated, to be effective in preventing economic gridlock, government deficit spending must be directed toward utterly useless production—industrial output that will not house, feed, clothe, or otherwise benefit anybody in any way. It must, in short, be deliberately wasted. Despite empirical evidence that government spending on "health care, education, clean energy, and infrastructure creates more jobs, dollar for dollar, than military spending,"[36] it does not lead to a *net increase* in employment. That is because using government money to construct schools, pay teachers, and build housing, highways, and bridges competes with private capital, which puts downward pressure on the number of jobs in the private sector and on the purchasing power they represent.

The principle of deliberate waste was first put on display in Rooseveltian public works programs that produced nothing at all—most notoriously exemplified by legions of workers with shovels digging long ditches to nowhere and then filling them back in again. As pointless as such activity would seem to be, it gave workers paychecks that allowed them to buy some of the surplus production without having them create more surplus products. But the apparent wastefulness was an insult to reason, and it was impossible in the American political context to explain that the paradox is an inherent feature of the capitalist economic system.

How could the irrationality of wasting social resources on such a grand scale be justified? The supreme boondoggle was found in weapons systems said to be necessary for national security. Thus was born the ever-increasing war budget, which has been the primary source of science funding ever since.

That is the full implication of the awkward phrase "weaponized Keynesianism." The US economy in its current form can survive only by flushing a never-ending, ever-growing torrent of money down the military toilet. The sad corollary is that the major portion of American science has been and still is being devoted to a vast exercise in intentional waste.

Well-Meaning but Missing the Point?

Well-meaning public-interest advocates frequently point out how the money lavished on the military could be better spent. Economist Joseph Stiglitz and public policy authority Linda Bilmes, for example, cited the $3 trillion the United States paid to wage the first few years of its war in Iraq and offered this comparison:

> A trillion dollars could have built 8 million additional housing units; could have hired some 15 million additional public school teachers for one year; could have paid for 120 million children to attend a year of Head Start; or insured 530 million children for healthcare for one year; or provided 43 million students with

four-year scholarships at public universities. Now multiply those numbers by three.[37]

These figures offer powerful evidence of how grotesquely deformed the priorities of American public policy have become. But they can be seriously misleading if they imply that the problem can be solved by simply redirecting war funding toward fulfilling social needs.

As previously noted, war spending is not primarily for war but to prevent economic gridlock, and socially beneficial spending cannot do that. To think runaway military budgets can be curbed by replacing warhawk politicians with progressives who will vote for a "more reasonable balance" between military and social spending is to miss the point.

If weaponized Keynesianism accurately describes the real nature of the American economy, the only remedy for its incurable addiction to military spending is cold-turkey abstention. That would require replacing economic institutions that base production decisions on maximizing the private wealth of the few with new ones that maximize the well-being of the many.

Opportunity Cost: The Paths Not Taken

The concept of opportunity cost in economics and accounting has to do with recognizing the value of alternative uses of resources. What might have been accomplished if the many trillions of dollars and incomprehensible magnitude of scientific talent wasted on military R&D had instead been devoted to worthwhile goals? It is impossible to know. Historian of science Stuart Leslie states the proposition this way:

> The full costs of mortgaging the nation's high technology policy to the Pentagon can be measured only by the lost opportunities to have done things differently. No one now can go back to the beginning of the Cold War and follow those paths not taken. No one can assert with any confidence exactly where a science and technology driven by other assumptions and priorities would have taken us.[38]

One nonetheless cannot help but believe that our species would be far, far better off if our collective efforts to accumulate scientific knowledge had focused, as political economist Seymour Melman put it, on "alleviating human wretchedness of long duration." With regard to the militarization of American science, "the true cost is measured by what has been foregone."[39]

Finally, neither the sheer wastefulness nor the lost opportunities represented by trillions of dollars in research and development of unused weapons is the worst part of the story. By the measure of human suffering, much more costly than bombs that have never been dropped are the ones that have: "Millions and millions of tons of bombs dropped; millions and millions of dead, mostly, of course, civilians."[40] The fruits of all that misbegotten science and technology will be the subject of the next seven chapters.

CHAPTER 12

A-Bombs and H-Bombs

WE MUST BE THANKFUL that most of the trillions of dollars in American military spending have been wasted. How much better off the world would be if *all* the vast mountain of money spent on military R&D, weapons manufacturing, and war preparations had been wasted. And how wonderful if we could be certain that the thermonuclear bombs will forever remain undropped.

Alas, the fruits of the military's science budgets have not all been tossed on the scrap heap. Many expensive high-tech war toys have proven irresistible to the generals and admirals who have used them to rain death and destruction in many parts of the world, most notably in Southeast Asia in the 1960s and '70s and the Islamic world from the 1990s to the present.

Where did all the money go? What was it spent on? What follows in this and the next six chapters is a partial catalog of what trillions of dollars of military research have bought us in the decades since World War II.

The catalog has to be partial because a comprehensive listing would be an unbearably tedious alphanumerical soup of B-52s, F-117s, W76-2s, and so on. I will therefore confine it to broad categories of the science-based weapons that pose the gravest dangers, that have caused the most harm to humankind, and that have devoured the most material resources. Among them are nukes, drones, and "antipersonnel" weapons of various descriptions.

The Nukes

The most dangerous and expensive category of high-tech weaponry has been, is, and will continue to be the nuclear arsenal.[1] The United States spent more than $9 trillion on nuclear weapons between 1940 and 1996,[2] producing an arsenal that grew from two bombs in 1945 to more than thirty thousand by 1965. After the Cold War ended, the number of nuclear warheads stockpiled or deployed began to decline; by 2019 it was down to thirty-eight hundred.[3] The existential threat they represent, however, has not diminished, if only because there are now at least eight other countries with nuclear capability.

While no A- or H-bombs have been dropped on populations since August 1945, that does not mean they have not been used. To the contrary, they have been used hundreds of times by several nations, mainly to intimidate other nations. The essence of deterrence strategy, after all, is intimidation. In August 2017, when Donald Trump threatened to unleash "fire and fury" against North Korea, that was but one in a long string of uses to which the most potent of the world's thermonuclear arsenals has been put.

During several decades of Cold War, the US and Soviet nuclear arsenals were also used to provide cover for a number of proxy wars between the two superpowers. The nuclear "balance of terror" allowed for the decimation of populations of smaller allies without causing annihilation of the primary combatants. The long years of US war in Southeast Asia claimed the lives of an estimated three to four million men, women, and children,[4] with nary a one killed by nuclear weapons.

Although undropped, the stockpiled nukes remain the primary focus of American militarized science policy and therefore deserve to top the list of high-tech weapons our tax dollars pay for. In 2004 antinuclear campaigner Helen Caldicott accurately described the US and Russian nuclear arsenals as they existed at that time:

> The U.S. currently has 2,000 intercontinental land-based hydrogen bombs, 3,456 nuclear weapons on submarines roaming the seas 15 minutes from their targets, and 1,750 nuclear weapons on intercontinental planes ready for delivery. Of these 7,206 weapons, roughly 2,500 remain on hair-trigger alert, ready to be launched at the press of a button. Russia has a similar number of strategic weapons, with approximately 2,000 on hair-trigger alert. In total

there is now enough explosive power in the combined nuclear arsenals of the world to "overkill" every person on earth 32 times.[5]

Despite their formidable stockpile, American policymakers soon came to perceive it as inadequate. As a result, just five years later president Barack Obama signed a National Defense Authorization Act to "modernize the nuclear weapons complex."[6]

The Obama Upgrade—"Better Nukes for Peace"?

The Obama administration thus committed to spending some $300 billion over the following ten years to upgrade and expand the American nuclear arsenal. In this endeavor, science and technology will be the starting point for all that follows.[7] In addition to replenishing the stockpile of nuclear warheads, the plan called for twelve new nuclear-capable submarines, one hundred new bombers, and four hundred new intercontinental ballistic missiles to deliver them. The $300 billion was just the initial ante; the thirty-year cost of the upgrade was projected to be well over $1 trillion.[8]

This major escalation of the arms race was carried out by a president who had won office after campaigning for a world without nuclear weapons. In April 2009, a few months after taking office, he began a major policy speech in Prague by alluding to the end of the Cold War and declaring:

> In a strange turn of history, the threat of global nuclear war has gone down, but the risk of a nuclear attack has gone up. More nations have acquired these weapons. Testing has continued. Black market trade in nuclear secrets and nuclear materials abound. The technology to build a bomb has spread.[9]

"As the only nuclear power to have used a nuclear weapon," he continued, "the United States has a moral responsibility to act." Therefore, he vowed, "I state clearly and with conviction America's commitment to seek the peace and security of a world without nuclear weapons."

Those were fine words, and for them, in October 2009, Barack Obama was awarded the Nobel Peace Prize. But just twenty days later, the new Nobel laureate signed into law the act initiating a *trillion-dollar expansion* of the US nuclear arsenal. In justification, he said he continued to seek disarmament but was obliged to negotiate from a position of strength, which required a reliable, modernized nuclear deterrent.

Obama's disarmament legacy was to turn over authorization for unprecedented expenditures on nuclear weapons to the warhawks of the Trump administration, who immediately dispensed with the pretense that it was actually in the interests of disarmament. General James Mattis, Trump's secretary of defense at the time, declared that "recapitalizing the nuclear weapons complex of laboratories and plants" was "long overdue."[10]

As could have been predicted, Obama's massive upgrade did not result in reductions in nuclear stockpiles around the world. Exactly the opposite happened: Trump's embrace of Obama's plan was perceived by Russia as raising the stakes in the nuclear poker game, to which Vladimir Putin replied by seeing Trump's raise and raising it again.[11]

"Little" Nukes—No Big Deal?

As of February 2018, the US arsenal included about five hundred relatively small "tactical" or "nonstrategic" nuclear weapons, two hundred of which were deployed with aircraft in Europe and the rest in the United States. High on the Trump administration's nuclear upgrade wish list was the creation of a new tactical warhead for submarine-launched missiles. That plan became reality in January 2019, when W76-2 warheads began rolling off production lines. A year later, in early 2020, they were attached to "an undisclosed number of Trident ballistic missiles" aboard US Navy submarines that are now roaming the world's oceans.[12]

The W76-2s are considerably less powerful than the old warheads they will replace—five kilotons of TNT equivalent as compared to one hundred kilotons. If that sounds like a step in the direction of reducing the threat of nuclear war, think again. In fact, they raise the danger precipitously. The hundred-kiloton behemoths that the Tridents presently carry are so enormous that their only possible utility is to deter a nuclear-armed enemy from daring to launch a first strike. The smaller W76-2 "isn't designed as a deterrent," historian James Carroll points out; "it's designed to be used."[13] The warhawks claim the "little" nukes could be routinely used in regional wars without triggering a massive, planet-destroying nuclear holocaust. That is "a fool's fantasy," Carroll explains, because "any use of such a weapon against a similarly armed adversary would likely ignite an inevitable chain of nuclear escalation whose end point is barely imaginable."[14]

The W76-2s are only small in the relative sense. Their power still amounts to a third of that of the bomb that obliterated Hiroshima. While the attempt by military policymakers to downplay the dangers of tactical nuclear weapons is designed to persuade the public that dropping them would be no big deal, in reality, one dropped on an urban area could be expected to kill tens of thousands of people.[15] And the possibility of a nuclear strike being confined to a single bomb is the least likely scenario imaginable.

From Davy Crockett to "Dial-a-Yield"

"The fixation with tactical nuclear weapons reached its apogee in the 1960s," arms control theorist Michael Krepon explains.[16] Herman Kahn hailed their usefulness in his seminal *On Escalation*, which argued that smaller nukes were necessary to create "gradations" in responses to threats from enemies. They would, Kahn contended, allow the generals to escalate wars while maintaining "escalation control."[17]

The pioneer of tactical nukes appeared in 1961, bearing the name of an American frontier hero with pop culture name recognition, Davy Crockett. The Davy Crockett was a portable bazooka that featured the smallest-yield nuclear warhead the Pentagon has ever created. A small squad of soldiers could tote the Davy Crockett around in the field, set it up on its tripod, and fire off its nuclear "cannonball" at will. Historian of science Alex Wellerstein summarized the device's career:

> The Davy Crockett system was actively deployed from 1961 through 1971. The redoubtable *Atomic Audit* reports that they were found to be highly inaccurate and were not effectively integrated into actual war plans. Nonetheless, according to the same source, some 2,100 warheads for the Davy Crockett system were produced, at a cost of about half a billion (1998) taxpayer dollars.[18]

The distinction between large strategic and small tactical weapons is blurred by the fact that some devices—most notably, the B61 bomb—can serve as both. The B61 is the primary thermonuclear gravity bomb in the US arsenal. It is a variable-strength ("Dial-a-Yield") bomb that can be set to yield anywhere from 0.3 to 340 kilotons, or from about twice the potency of the Davy Crockett to about twenty-two times that of the Hiroshima bomb.

Comparisons of explosive power don't tell the whole story of the nukes' lethality. Their killing capability is a function not only of the power of their blast but also of the long-term environmental poison they leave in their wake in the form of lethal radiation. In Hiroshima and Nagasaki, tens of thousands of people died from radiation poisoning in the months and years after the atomic bombs were dropped.

The production of the new W76-2 warheads has taken us across an extremely frightening line. It brings the possibility of nuclear warfare closer than at any time since the end of the Cold War.

CHAPTER 13

Non-nuclear Technologies of Death

DURING THE VIETNAM WAR, antiwar activists frequently claimed that the United States had dropped more bombs on economically less-developed Southeast Asian countries than had been dropped on all of industrialized Europe throughout World War II. That was a powerful condemnation of US war policy . . . but was it true?

Although it sounds exaggerated, it was in fact *under*stated. Reliable records attest that 7,662,000 tons of bombs were dropped by the United States in Southeast Asia, compared to less than a third of that amount—2,150,000 tons—dropped in Europe between 1939 and 1945.[1]

The end of the Vietnam War did not mean an end to bombing, although its focus shifted from Southeast Asia to other parts of the world. After the September 11, 2001, attacks, the Bush-Cheney administration launched a ferocious air war against targets in the Middle East. By the end of March 2002, some 20,000 US bombs had been dropped on Afghanistan. In March 2003, the United States began its invasion of Iraq with the infamous "shock and awe" air assault on Baghdad. In the following weeks, 863 American planes conducted 24,000 bombing runs that left an estimated 2,700 civilians dead.[2] Human Rights Watch reported that the operation utilized "almost 13,000 cluster munitions, containing nearly 2 million submunitions, that killed or wounded more than 1,000 civilians."[3]

As the years passed, the bombing of the Middle East continued unabated. Between 2009 and 2012, 18,000 bombs were dropped or delivered by drone or missile in Afghanistan.[4] And Pentagon figures state that the United States and its allies conducted 24,566 air strikes in Iraq and Syria from August 2014 through August 2017.[5]

In 2016, Obama's final year as president, the number of bombs dropped was conservatively estimated at 26,171.[6] With Trump and his generals in the White House, the pace quickened. In addition to "unprecedented levels" of bombing in Iraq and Syria, "in Afghanistan, the number of weapons released has also shot up since Trump took office." Furthermore,

> Trump has also escalated U.S. military involvement in non-battlefield settings—namely Yemen, Somalia, and Pakistan. In the last 193 days of the Obama presidency, there were 21 lethal counterterrorism operations across these three countries. Trump has quintupled that number, conducting at least 92 such operations in Yemen, seven in Somalia, and four in Pakistan.
>
> Increased air power in Iraq and Syria has resulted in unprecedented levels of civilian deaths. Even by the military's own count, civilian casualties have soared since Trump took office, though independent monitors tally the deaths as many as ten times higher. In Afghanistan, Trump's tolerance for killing civilians has led to 67 percent more civilian casualties in his first six months than in the first half of 2016, according to the United Nations.[7]

Antipersonnel Bombs

What, exactly, is an "antipersonnel bomb"? Both parts of that term require closer examination.

In everyday parlance, the word "personnel" designates an indeterminate number of faceless employees. In the same spirit, military lingo uses "antipersonnel" to depersonalize the victims of weapons designed to kill large numbers of human beings.

The word "bomb" conjures up an image of a simple device dropped from an airplane that does its damage by exploding when it hits the ground. But that tends to undervalue the contributions of American science to the technological wizardry of modern warfare. Some bombs drop, but many are guided in sophisticated ways toward their targets. And bombs no longer merely explode; they have been engineered to accomplish their antipersonnel mission in more devious ways. Some incinerate rather than blow up their victims, and others incorporate clever ways of scattering their killing power over vast areas.

Napalm

Napalm has, for good reason, become an enduring symbol of inhumane and immoral weapons of war. A jelly mixture designed to adhere to human skin and melt the flesh, napalm causes burns far more severe than fire, and wounds too deep to heal. The appalling cruelty of the injury and death caused by napalm was revealed to the American public via "the televised war" in Vietnam. Mass protests erupted against the chemical weapon's use by the military and its manufacture by the Dow Chemical Company.

Napalm originated during World War II in the Harvard University laboratories of a team of scientists headed by organic chemist Louis F. Fieser. Their assignment was to upgrade the lethality of existing incendiary weapons.[8] World War I flamethrowers shot out flaming gasoline, but it burned out too quickly to cause much damage. Later, rubber was mixed with the gasoline to remedy that shortcoming. As an added bonus, the new mixture would also cling to its victims' skin, thereby intensifying its "antipersonnel" effectiveness.

With rubber in short supply when the US entered World War II, the Pentagon turned to the scientists, and in January 1942 napalm was born. Napalm made its combat debut in mid-1942, and by the end of the war, US military use amounted to seventy-five million pounds a year.

Before the atomic bombings of August 1945, US forces firebombed more than sixty of Japan's largest cities, killing more people than those who perished at Hiroshima and Nagasaki. The US Strategic Bombing Survey estimated that 157,000 tons of bombs were dropped on Japanese population centers between January 1944 and August 1945, killing 213,000 and leaving fifteen million homeless. The March 1945 firebombing of Tokyo was especially atrocious. With characteristic callousness, General Curtis LeMay later boasted that the attack, which he had directed, "scorched and boiled and baked to death more people in Tokyo on that night of March 9–10 than went up in vapor at Hiroshima and Nagasaki combined."[9]

That was the prelude to what was to come in Vietnam, where LeMay famously announced his intention "to bomb them into the Stone Age."[10] Although that ambition was not realized, it was not for lack of effort. An estimated 388,000 tons of napalm were dropped on Vietnam between 1963 and 1973.[11] Due also to the more than twenty million gallons of herbicides such as Agent Orange that were sprayed on Vietnam, chemical warfare left a legacy of environmental devastation, cancer, and birth defects, with an impact still felt today.[12]

The worldwide revulsion sparked by Vietnam's experience with napalm led the United Nations General Assembly, in 1980, to adopt a treaty outlawing its use on civilians. The United States did not ratify the treaty until twenty-nine years later, in 2009, and then only with a "reservation" saying, in essence, that the United States "reserves the right" to ignore the prohibition whenever it wants. In fact, the US military has used napalm bombs under aliases like "Mark 77" in recent years in Iraq and Afghanistan, and has kept them in its arsenal to the present.

In 2003 vociferous Pentagon denials that napalm was being used in Iraq were exposed as untrue. Confronted with evidence of the falsehood, a DoD spokesperson drew a distinction between napalm bombs and Mark 77 firebombs, acknowledged that the latter had been used, and admitted that they were "remarkably similar" to napalm weapons. Robert Musil of Physicians for Social Responsibility called distinguishing between napalm and Mark 77s "outrageous" and "Orwellian."[13]

Cluster Bombs

Napalm's reputation as a heinous weapon is well deserved, but it did not kill and maim as many people as another great advance in the technology of death, the cluster bomb. The first cluster bombs were developed independently during World War II by Germany, Italy, the Soviet Union, and the United States. The technology was not sophisticated; in the American version, a dozen or so twenty-pound fragmentation bombs were simply wired together.

The modern version that evolved from that prototype can contain hundreds—sometimes thousands—of "submunitions," each about the size of a soda can or a tennis ball. Cluster bombs are designed to explode in midair, scattering their bomblets across areas of up to twenty acres. The bomblets blow up on contact and eject shrapnel. One exploding bomblet can kill or wound anyone within a radius of a hundred yards.

Bombing during the Vietnam War was not limited to Vietnam; a great deal of it was inflicted upon neighboring Laos and Cambodia. From October 1965 through May 1975, at least 456,365 cluster bombs were dropped on the three Southeast Asian countries.[14] The Pentagon could claim that bombing Vietnam was legal because it was carried out with the blessing of the South Vietnamese government. The Laotian and Cambodian governments were officially neutral, however, so by recognized standards of international law, American bombing of those countries was illegal. US officials privately debated how best to present it to the world as lawful, but were committed to the bombing regardless of its legality.

A Decade of Atrocity in Laos

In the annals of unjust aggression and "asymmetric warfare," Mussolini's pre–World War II invasion of Ethiopia was notorious for sending machine guns, warplanes, and poison gas against the Abyssinians' arrows and spears. But the US air war on Laos in the 1960s and '70s was no less one-sided and far more murderous. The most formidable military power on earth dropped *two and a half million tons of explosives* on a tiny, impoverished, defenseless country—the equivalent of a planeload of bombs every eight minutes, twenty-four hours a day, for nine years.[15]

The assault on Laos was directed by the CIA rather than the Pentagon. Called "the secret war" and "the largest covert action in American history," it was an open secret, hidden from no one. The CIA's role in Laos marked a major step in militarizing the spy agency, which now routinely carries out paramilitary operations and drone strikes in many parts of the globe.[16]

A US Congressional Research Service report described the extent of the bombing:

> Laos has been characterized as the most heavily bombed country in history, on a per capita basis. From 1964 through 1973, the United States flew 580,000 bombing runs over Laos and dropped more than 2 million tons of ordnance on the countryside, double the amount dropped on Germany during World War II.[17]

That tonnage comprised 270 million cluster munitions.[18] The population of Laos was about three million, which means that ninety lethal explosives were indiscriminately dropped on Laos for every man, woman, and child in the country.

Although most of the bomblets exploded as intended, a significant number—up to 30 percent—did not. They constitute the dreaded UXO—unexploded ordnance. To this day, an estimated eighty million bomblets remain scattered throughout rural Laos, killing, blinding, and blowing off limbs.[19]

UXO—The Gift That Keeps on Giving

Since the aerial bombardment ended in 1973, UXO has caused more than twenty thousand deaths or grievous injuries in Laos.[20] Children are particularly endangered, because they often mistake the small bomblets for toys.[21] Laotian farmers risk their lives every time they plow a field or push a shovel into the soil, but they have few options. If they don't farm their land, they starve.

Why Laos? What possible motive could have justified such an extreme and barbaric course of action? US military and government spokespersons offered only one explanation: they were afraid Laos would "go Communist."

In the end, Laos did indeed go Communist, but no Laotian threat to US "national security" ever materialized. In fact, American policymakers not only learned to live with a Communist Laos but also worked to integrate it into their economic and diplomatic world order. In the late 1970s, they turned to diplomatic means to regain influence (and counter rising Chinese influence) in the region.

In its pursuit of a "strategic partnership" with Laos, the United States contributed $100 million over twenty years to UXO cleanup efforts.[22] In September 2016, Barack Obama became the first US president to visit Laos and pledged another $90 million over the following three years to continue the program.[23]

Obama explicitly *did not* offer an official American apology to Laos for dropping the cluster bombs in the first place, asserting that the United States had been on the "right side of history" during the Cold War. Nevertheless, he acknowledged a US "moral obligation to help Laos heal," and—ever the master of understatement—conceded that the bombing might not have been the best way to win over the Lao people's hearts and minds.[24]

After Laos, the US military continued to use cluster munitions, most notably in the wars against Afghanistan and Iraq.[25] In 2008 an international Convention on Cluster Munitions was proclaimed, banning the use and stockpiling of cluster bombs. As of March 2018, 103 nations had acceded to the treaty, but the United States was not among them. In response to the ban, George W. Bush, just before leaving office in 2008, pledged to "sharply restrict" US use of cluster bombs. Meanwhile, a State Department official appealed to science to justify continued American commitment to the weapons:

> Like it or not, the rules of physics and chemistry dictate that cluster munitions are the most effective conventional means to destroy many types of military targets spread over an area. . . . Cluster munitions are available for use by every combat aircraft in the U.S. inventory, they are integral to every Army or Marine maneuver element and in some cases constitute up to 50% of tactical indirect fire support. U.S. forces simply can not fight by design or by doctrine without holding out at least the possibility of using cluster munitions.[26]

Despite Bush's pledge, throughout Barack Obama's eight years in office the Pentagon retained a stockpile of more than two million cluster bombs in the United States and another million and a half in South Korea. In 2009 navy warships fired Tomahawk cruise missiles carrying cluster munitions at Yemen, and in 2015 the United States supplied cluster bombs to its Saudi allies to drop on Yemeni villages and towns.

All pretense of "sharply restricting" cluster munition use finally came to an end in December 2017, when the Trump administration officially rescinded Bush's 2008 pledge. "After spending hundreds of millions of dollars researching alternatives to cluster munitions," Human Rights Watch declared, "the US has decided it can't produce 'safe' cluster munitions so it will keep using 'unsafe' ones."[27] The US Army has already begun conducting "lethality arena tests" on "the M999 advanced cluster munition" in Israel.[28]

CHAPTER 14

Bombers, Missiles, and Antimissiles

BOMBS ARE ONLY USEFUL as weapons of war to the extent that they can be delivered to the warriors' desired targets. Of the research funds devoted to nuclear warfare from 1940 to 1996, 57 percent was spent on developing delivery methods as opposed to 7 percent on the weapons themselves.[1]

The original delivery option—long-range bombers to carry loads of explosives to distant targets—continues to be improved for the task. Within a year of the end of World War II, the first bomber with the specific purpose of carrying nuclear payloads—the B-36—was produced by the Convair Corporation. In the mid-to-late 1950s the B-36 was eclipsed by Boeing's more versatile B-52, which is still in use and is projected to remain in active service into the 2050s. In the Vietnam War era, the B-52 design was modified to allow it to carry up to sixty thousand pounds of bombs.

How High—or Low—Can You Go?

On May Day 1960, the Soviet Union shocked the world by downing an American U-2 surveillance plane with a surface-to-air missile and capturing its pilot, Gary Powers. The U-2, which could fly more than thirteen miles above the earth, had previously been considered untouchable, but from that moment forward the vulnerability of aircraft at the highest altitudes had to be taken into account.

The US Air Force had already begun research into the science of enabling planes to fly as low as possible instead of as high as possible, but the U-2 incident gave it new urgency. By hugging the earth's surface, low-flying

168

aircraft were better able to evade radar detection and sneak up on their targets, giving rise to the familiar metaphor "flying under the radar."

Stealth Technology

The next generation of military researchers, in the 1970s, pursued novel technologies for deflecting or absorbing radar signals that would result in "stealth" aircraft far less visible to radar. The combination of stealth with aerodynamic stability had to wait for advances in the computerization of flight control, but by the early 1980s Lockheed's stealth fighter, the F-117 Nighthawk, had made its debut.

In 1991 an F-117 dropped the first bomb in the initial Operation Desert Storm air raid on Baghdad, followed by waves of F-117s that pounded the city's densely populated downtown area. The Iraqi government reported that the air strikes killed twenty-three hundred civilians. That none of the F-117s were shot down was deemed empirical evidence supporting the scientific principles underlying stealth technology. F-117s went on to fly a total of 1,280 Desert Storm combat sorties in the First Gulf War.

The second aircraft with advanced stealth capability was the B-2 bomber, originally designed to carry nuclear bombs. Converted for use with conventional bombs, it first saw combat action in the Kosovo War in 1999, and continued its career in Iraq, Afghanistan, and Libya. B-2s dropped more than a million and a half pounds of bombs during the 2003 US air assault on Iraq with the Orwellian code name "Operation Iraqi Freedom."

Stealth bombers are expensive. A 1997 US Government Accounting Office report estimated the total B-2 program cost at more than $2 billion *per aircraft*.[2] On top of that, the B-2 costs an estimated $130,000 an hour to fly.[3] As of February 2018, the air force was pursuing a major modernization of the B-2 while readying a next-generation stealth bomber, the B-21, to take its place. The Obama-Trump nuclear upgrade aims to add another ultra-high-tech aircraft to the mix: the F-35 Joint Strike Fighter.

Is the F-35 "Too Big to Fail"?

At a projected total cost of almost $1.5 trillion ($1,450,000,000,000), the F-35 is the most expensive weapons system in the history of the world. A 2019 assessment puts that price tag in perspective:

Development and procurement of roughly 2,400 F-35s through 2037 is now estimated to cost over $400 billion, roughly *eight times the annual defense spending of Russia*. Operating expenses through 2070 are estimated to cost an additional $1.1 trillion.[4]

The concept of the F-35 originated in 1992, the final year of the G. H. W. Bush administration. It emerged from a combined air force and navy project dubbed the Joint Strike Fighter program, but the high-tech aircraft only began to take material form in 2001, when the Pentagon awarded a contract to create it to Lockheed Martin.

The F-35 offered dazzling, ahead-of-the-curve technological wizardry. One of the warplanes' advocates hailed them as "flying computers" that could "detect an enemy plane five to 10 times faster than the enemy could detect it."[5]

Six infrared cameras mounted on the exterior stream real-time imagery, allowing the pilot to "see through" the skin of the plane, including straight down. In a process called "sensor fusion," the main onboard computer can meld data from the exterior cameras, the plane's powerful radar, and an "electro-optical" targeting system.[6]

This 360-degree view of what's outside the plane is projected onto the curved visor of the pilot's custom-built helmet. The price of these virtual-reality headsets is half a million dollars . . . *each*. But futuristic technology is only valuable insofar as it actually works. Nor can it stay "ahead of the curve" if it isn't brought into being in a timely fashion. The first flight of an F-35 took place in 2006, and Lockheed began to deliver the planes to the air force in 2007, but they were not the finished product that had been promised. They were plagued with innumerable major and minor design flaws, incurring new costs and requiring lengthy delays to fix.

In 2011, senator John McCain of the Senate Armed Services Committee blasted the F-35 program as "both a scandal and a tragedy."[7] Vermont senator Bernie Sanders condemned the warplane for its "massive cost overruns that have wasted hundreds of billions of taxpayer dollars."[8] In 2018 the F-35 entered its seventeenth year of continual cost overruns and production delays, and the finish line still remained in the hazy distance. Despite the congressional huffing and puffing, however, the project is not and never was in jeopardy. By the end of 2018, Lockheed

Martin was boasting that "more than 355 F-35s have been delivered and are now operating from 16 bases worldwide."⁹ Earlier, when a squadron of F-35s was stationed at a marine corps base in Arizona, Senator McCain's home state, he changed his tune. The F-35, he gushed, "may be the greatest combat aircraft in the history of the world."¹⁰

Meanwhile, Bernie Sanders experienced a similar epiphany and supported a proposal to equip the Vermont Air National Guard with eighteen F-35s. When challenged by protesters at a town hall meeting, he testily responded:

> My view is that given the reality of the damn plane, I'd rather it come to Vermont than to South Carolina. And that's what the Vermont National Guard wants, and that means hundreds of jobs in my city. That's it.¹¹

That, however, is only a small part of why the F-35 is Too Big to Fail. The warplanes, one analyst explains,

> require a total of some 300,000 parts, and Lockheed has parceled out the subcontracting to all but five unlucky states—Alaska, Hawaii, Louisiana, North Dakota, and Wyoming. Lockheed says the F-35 directly or indirectly supports 146,000 jobs across the U.S., ranging from minimum-wage broom-pushers to engineers paid well into six figures.¹²

The Pentagon and its political enablers have no intention of backing off from their trillion-and-a-half-dollar commitment to the F-35. To most observers, including many who share the military ethos, devoting that much money to weapons systems of questionable utility may seem irrational, but war spending has a logic all its own.¹³

In the meantime, the rapid improvement of air defense technology had led R&D resources to flow toward ballistic and guided missiles to which nuclear and conventional warheads could be attached. Missiles fly ten to twenty times faster than the fastest aircraft and can go places and do things long-range bombers, and even stealth fighters, can't. Missiles did not, however, make bombers obsolete; they complemented each other. B-52s were modified, and newer aircraft were designed, to carry and fire cruise missiles.

Rocket Science: ICBMs, MIRVs, and Cruise Missiles

The common phrase "It ain't rocket science" expresses a general awareness that missile technology proceeds from scientific praxis of the highest order. Propelling a heavy warhead skyward is fairly simple, but having it come down at a faraway predetermined location is quite a sophisticated feat.

Ballistic missiles are blasted off with great force, but the rest of their flight, following a preplanned trajectory, is unpowered. The distance from launch to target typically ranges from a few hundred to several thousand miles, and they can travel at speeds of up to fifteen thousand miles per hour. The longest-range of the genre, the Intercontinental Ballistic Missiles, made their debut in the late 1950s. The Soviet Union launched the first test flight of an ICBM, the Semyorka rocket, on August 21, 1957. Its first operational ICBM followed in 1958, and the United States' initial entry into the competition occurred soon thereafter, in 1959.[14]

ICBM trajectories take them far above Earth's atmosphere into suborbital space flight, reaching altitudes as high as twelve hundred miles before plunging earthward toward their targets. They can reach any spot on Earth from any launch site. North Korea, for example, is believed to possess ICBMs capable of traversing the 6,830 miles from liftoff to Washington, DC, in an estimated 40 minutes and 30 seconds. But the most spectacular of rocket science's achievements is the accuracy with which its missiles can strike.

After traveling thousands of miles with no external guidance system controlling its path, an ICBM can reasonably be expected to land within a few hundred feet of its intended target. The US arsenal's Minuteman III and Trident II missiles will on average miss their marks by no more than seven hundred feet and four hundred feet, respectively. That means that most of a fleet of nuclear-armed ICBMs aimed at major cities would fall close enough to the cities' centers to assure their total destruction.

The system of land-based ICBMs that the Trump administration inherited from its predecessors consists of 399 Minuteman III missiles, ensconced in heavily fortified underground silos scattered throughout the western United States.[15] They are sixty feet long and five and a half feet in diameter, and they weigh seventy-eight thousand pounds each. That these missiles are on "hair-trigger alert" is a common cliché, but there is no reason to doubt it. Once fired, they cannot be recalled.

This entire fleet of ICBMs, however, is to be scrapped as part of the Obama-Trump nuclear upgrade plans. The 399 Minuteman IIIs are to be replaced by a system of 400 new missiles designated by a suitably bureaucratic set of initials, GBSD ("Ground Based Strategic Deterrent").

The ICBMs' past, present, and future cost to taxpayers is shrouded in official secrecy, but we have a credible estimate from General John Hyten, head of the US Strategic Command.[16] According to General Hyten, the original ICBM program that began in the late 1950s provided 800 Minuteman I missiles for the total price of $17 billion (in 2017 dollars). By contrast, the 400 new missiles of the proposed GBSD program are expected to cost $85 billion. In other words, the unit price of the Minuteman I missiles was about $20 million, whereas their new counterparts, at about $200 million apiece, will be ten times as expensive.[17]

MIRVs—More Bangs per Bomb

ICBMs became considerably more lethal when the MIRV—Multiple Independently-targeted Reentry Vehicle—was introduced in the late 1960s. A RAND Corporation scientist who played a leading role in creating the ICBM, Bruno Augenstein, has also been credited with inventing the MIRV. Taking advantage of progress in the miniaturization of nuclear warheads, he devised a way to fit several on a single missile.[18] The warheads of an MIRV can be separately aimed at multiple targets.

The enhanced efficiency of the MIRV, however, turned out to be problematic for US policymakers. The problem wasn't with the science or technology but with the international political situation. MIRVs dangerously destabilize the standoff that Mutually Assured Destruction is supposed to provide.[19] When one country can wipe out many of its enemy's warheads with a single attack, it increases the "use 'em or lose 'em" incentive for preemptive first strikes.

Despite that risk, both the United States and the Soviet Union stubbornly maintained large numbers of MIRVs in their arsenals throughout the Cold War. Finally, between 2010 and 2014, the United States dismantled its land-based MIRVs to reduce the likelihood of attracting a surprise attack. Its submarine-based MIRVs were retained, however, because the submarines' mobility and beneath-the-surface invisibility are believed to make them invulnerable to attack. As of 2017, the US Navy had fourteen

"Ohio-class" submarines armed with MIRVed Trident II ballistic missiles. Each of the fourteen subs carries twenty-four Tridents and each Trident contains four warheads, each several times more powerful than the Hiroshima bomb. That means the United States has 1,152 thermonuclear weapons deployed and ready to launch from ever-changing locations throughout the oceans of the world.[20] And that doesn't include the cruise-missile nukes.

Cruise Missiles

Cruise missiles are unmanned, jet-powered, guided projectiles that can be launched from land, air, or sea. Unlike ballistic missiles, they are not confined to parabolic trajectories but instead follow complex preprogrammed flight paths supplemented by in-flight external data and remote control. They can therefore actively maneuver to avoid antimissile attacks, and their external "pilots" can home in on targets seen via cameras mounted in the missiles' noses.

Ballistic missiles employ inertial-guidance technology, which uses data from accelerometers and rotation detectors (gyroscopes) to estimate their position by the classic guesswork technique known as dead reckoning. Cruise missiles use inertial guidance, too, but they also receive data inputs from laser-based terrain contour mapping and satellite-based global positioning systems (GPS). The latter has escaped the confines of military technology to become a familiar staple of everyday civilian life.

Some cruise missiles, moreover, have the more advanced ability to track and attack moving targets utilizing radar-homing and infrared-homing (a.k.a. "heat-seeking") technology.

The Tomahawk Missile

To name and describe all of the cruise missiles in past and current use would require a very large volume, but the versatile Tomahawk missile, to which "America's military is addicted,"[21] can serve as an archetype. The US Navy is in charge of the Tomahawks, which can carry either nuclear or conventional warheads. All of the navy's cruisers, destroyers, and attack submarines are outfitted with Tomahawk weapons systems.

Tomahawk missiles, with an official price tag of about $1.9 million each, utilize preset guidance technology to follow a programmed course to a known, stationary land target. They can, however, be reprogrammed in midflight to change targets, and they can also "loiter" over battlefields awaiting redirection to more critical targets.

The Mystique of the Pinpoint-Precision Surgical Strike

All of these dazzling, innovative, and very expensive technologies have contributed to the widespread perception that today's "smart" missiles and bombs are capable of virtually perfect accuracy. That claim is actively promoted by the Tomahawk's manufacturer, the Raytheon Company:

> Today's Tomahawk Block IV cruise missile can circle for hours, shift course instantly on command and beam a picture of its target to controllers halfway around the world before striking with pinpoint accuracy.[22]

But as impressive as cruise missile guidance technology is, and despite the great battlefield advantage it provides, its accuracy has been greatly exaggerated to serve dishonest political agendas. The Pentagon's constant assurances that their air strikes are smart, surgical, and pinpoint-precise are the centerpiece of a PR campaign to downplay the extent of civilian casualties their weapons cause.

There are a number of problems with the pinpoint-precision scenario.

First is the "swatting a fly with a sledgehammer" problem: what is the value of hitting a pinpoint target with a bomb that destroys a city block?

Second is the "friendly fire" problem: what is the value of precise targeting if you don't know what you're aiming at? The Patriot missile defense system shot down a number of American and British fighter jets in Iraq because, as a Federation of American Scientists analysis reported: "The Patriot's Identify Friend or Foe (IFF) algorithms (which ought to have clearly distinguished allies from enemies) performed poorly."[23] Friendly fire incidents—accidental attacks on one's own troops or allies—have risen dramatically since World War II:

> According to the most comprehensive survey of casualties (both fatal and nonfatal), 21 percent of the casualties in World War II were attributable to friendly fire, 39 percent of the casualties in Vietnam, and 52 percent of the casualties in the first Gulf War.[24]

Third, and most fundamental, is the validity problem: the truth content of the military-industrial complex's pinpoint-precision claims cannot simply be taken for granted. In fact, evidence abounds that the missiles' guidance systems often malfunction and have serious shortcomings even when they function correctly.

The accuracy of GPS-guided weapons is undermined by their vulnerability to electronic jamming and cyberwarfare hacking. An even more basic limitation is GPS dependence on knowing where the target is in the first place. If inaccurate intelligence reports cause incorrect GPS coordinates to be programmed into the missile, unintended consequences are guaranteed. Faulty intelligence has been blamed for the deaths of more than a thousand civilians as "collateral damage" of the Obama administration's targeted killings in Afghanistan.[25]

Laser-guided weapons home in on their targets by following laser beams. Those beams, however, become unreliable in smoky, dusty, cloudy, or rainy conditions, because particles in the air reflect and refract laser light. Common weather conditions, or the smoke and dust created by previous explosions, can cause the missiles to miss their targets.

Fourth is a problem that might be called the "big picture disconnect." If pinpoint-precision weapons are the solution to the problem of civilian deaths, why do vast numbers of civilians—more than ever in the smart-bomb era—keep dying? Can it be that the bombs' smartness in fact serves mainly to increase the efficiency of their ability to kill civilians? Bamboozling the American public into thinking that precision bombing can avoid hitting noncombatants leads military commanders to bomb civilian areas they were previously forbidden to target. The end result is more, not fewer, civilian casualties. Despite the Pentagon's protestations to the contrary, one can reasonably suspect its officials of not really caring much about how many civilians of all ages are slaughtered by their bombs and missiles. Blunt-speaking General Curtis LeMay provided this classic expression of how the military mind frames the problem:

> There are no innocent civilians. It is their government and you are fighting a people, you are not trying to fight an armed force anymore. So it doesn't bother me so much to be killing the so-called innocent bystanders.[26]

Why Smart Missiles Are Less Smart Than Advertised

The Pentagon would rather you not know how dumb its smart weapons really are. In keeping that secret, they have the law on their side. The pinpoint-precision mystique is sustained by hiding evidence to the contrary behind the veil of official secrecy. Honest evaluations of the weapons' accuracy, either in tests or in combat, are few and far between. The most thorough peek behind the curtain came in a 1997 report by the Government Accountability Office on how well the missiles and other high-tech weapons performed during Operation Desert Storm.

The GAO's four-year study found a "pattern of overstatement" propagated by the weapons industry and echoed by Pentagon officials. In particular, it debunked the claim that Patriot missiles can intercept incoming weapons on a one-to-one basis.[27] In fact, to hit a single incoming rocket, Patriot missiles have to be fired off in swarms, a technique called "ripple fire." As one critic observed, the Patriot, although promoted as a precision weapon, "is actually just an expensive, high-tech version of buckshot."[28] "Expensive" is an understatement; each Patriot missile costs $3 million.

The GAO's revelations were later underscored by a few spectacular examples of smart-weapon stupidity that could not be hidden from public view. In 1999, during the Kosovo War, American GPS-guided bombs erroneously hit the Chinese embassy in Belgrade. During the 2003 assault on Baghdad, at least three American missiles not only missed the city, *they missed the entire country of Iraq!* On March 21, a US missile hit a building in Iran; on March 24, another one killed five civilians riding in a bus in Syria; and on April 8, yet another American missile hit Iran again, killing a thirteen-year-old boy.

Faster, Ever Faster: The Arms Race Speeds Up

A radically new technology, hypersonic weaponry, has become a central focus of the current arms race between the United States, China, and Russia. Missiles are categorized as hypersonic if they travel at or above Mach 5 (3,834.6 miles per hour, five times the speed of sound). Innovative "scramjet" technology lightens hypersonic cruise missiles by relieving them of the need to carry liquid oxygen tanks. Instead, a fanless engine "uses shock waves created by its speed to compress incoming air in a short funnel and ignite it while passing by."[29]

These missiles travel so fast that they don't even need to carry warheads to cause massive destruction. The kinetic energy of a five-hundred-pound missile slamming into its target at thousands of miles per hour is equivalent to many tons of high explosives. Nonetheless, the missiles are being designed to carry both conventional and nuclear warheads. "Hypersonic missiles are a game-changer," declares military analyst Steven Simon. "No existing defenses, in the United States or elsewhere, can intercept a missile that can move so fast while maneuvering unpredictably."[30]

Deterrence strategists warn that hypersonic missiles will seriously destabilize confrontations between hostile nations. These missiles "arrive at their targets in a blinding, destructive flash, before any sonic booms or other meaningful warning." That "will dangerously compress the time during which military officials and their political leaders" can decide how to respond.[31] Not knowing whether missiles flying toward them are nuclear-tipped or not, they could understandably decide to fire off nuclear warheads at their attackers.

In the last week of 2019, Russia announced that its Avangard missile—"a new intercontinental weapon that can fly 27 times the speed of sound" and "carries a nuclear weapon up to 2 tons"—had become operational and "had entered combat duty."[32] Furthermore, "analysts say the Chinese are even further along than the Russians" in developing hypersonic missiles.[33]

Meanwhile, the US military plans to have a hypersonic weapon operational by 2022 and the American military industry is salivating over the lucrative contracts it will generate. Lockheed Martin was awarded $1.4 billion in 2018 to build hypersonic missile prototypes, and in 2019 tested one that allegedly reached Mach 20. That's not quite "faster than a speeding bullet," but it's pretty close.

What about Drones?

Another major innovation in missile-delivery technology burst into public awareness in recent years: the now-ubiquitous Unmanned Aerial Vehicles (UAVs) commonly known as drones. Their impact on warfare has been so revolutionary as to require separate treatment: see chapters 15 and 16.

Missile Defense Systems: Porous Shields and False Security

With the thermonuclear-tipped ICBM, science bequeathed to humanity the ultimate sword for which there is no shield—a weapon against which there is no defense. In lieu of genuine protection, policymakers offered, first, the disingenuous illusions of deterrence strategy—the paradoxical notion that amassing a nuclear arsenal that could obliterate the planet many times over could guarantee our safety. That was complemented by "shields in the sky"—electronic missile defense systems to detect and destroy incoming ICBMs. The shields these schemes provided, culminating in the Reagan administration's Star Wars program, were at best highly porous. Rather than genuine protection, they promote a false sense of security.

From DEW Line to BMEWS

Before there were ICBMs, the threat of long-distance bombers bearing nukes haunted the minds of American Cold Warriors. They decided a chain of radar stations was needed to detect Soviet aircraft coming over the North Pole in time to shoot them down or, failing that, launch a counterstrike before the incoming bombs hit. And so in 1957 the Distant Early Warning Line—more snappily known as the DEW Line—was born. It consisted of fifty-seven radar stations spanning the North American continent, a hundred miles or so north of the Arctic Circle.

A history of the DEW Line prepared by the Air Defense Command acknowledged that in tests, the DEW Line's alarm system "triggered more false alarms than actual ones." The false alarms were attributed to "cloud formations, ice flows [sic] and electronic interference, among other things."[34] Those "other things" included meteor showers and flocks of Canadian geese.

Rather than *preventing* a danger, the DEW Line *presented* one. The system's propensity to false alarms raised the probability that the air force would erroneously launch an attack that would initiate a genuine nuclear exchange.

When the Soviet Union introduced ICBMs into its arsenal, the DEW Line was rendered obsolete, but rather than scrapping it, the Pentagon supplemented it with the more advanced BMEWS, Ballistic Missile Early Warning System. The concept of a chain of connected radar stations is a simple one, but actually creating it required, and stimulated, a great deal of scientific innovation. The conventional spinning radar antennas of World War II could not turn fast enough to track multiple ICBMs. In 1959 the Pentagon's premier scientific agency, DARPA,[35] offered a solution: phased array antenna systems that remained in fixed positions and could detect objects without being physically pointed at them.

Another 1959 scientific milestone was Bell Labs' invention of the MOS (metal-oxide-silicon) transistor, which revolutionized solid-state electronics. That, in turn, liberated computers from the constraints of vacuum-tube technology, vastly expanding their computing power while shrinking their physical dimensions. Essential advances in automating missile detection functions would otherwise have been inconceivable.

Construction started on BMEWS (pronounced "bemuse") in 1958, and the system began operations in 1961. Even before its official debut, however, preliminary tests at its Thule, Greenland, station exposed a fundamental flaw of missile defense systems in general. On October 5, 1960, the BMEWS station at Thule detected a large object rising on the horizon and identified it as a Soviet missile. An urgent warning was immediately sent to NORAD, the North American Air Defense Command, that enemy missiles were hurtling toward the United States at supersonic speed. The president, or someone acting in his stead, had only about ten minutes to decide whether or not to launch an immediate, all-out nuclear counterattack.

Fortunately, it was a false alarm, and the error was promptly detected. It so happened that Soviet premier Nikita Khrushchev was in New York City at the time, making it highly improbable that the USSR had actually launched a nuclear strike against the United States. That warranted a closer look at what the radar had seen rising on the horizon. It was the moon.

BMEWS and its successors would not likely make that particular mistake again, but it was a dramatic reminder of Murphy's Law: "Anything that can go wrong may well do so." Appropriately, the eponymous author of the maxim, Edward R. Murphy, was an aerospace engineer working on a US Air Force project. The greater the consequences of something going wrong, the more Murphy's Law should be heeded, and no consequence could be greater than accidentally starting a nuclear war.

But the lesson underlying the misidentification of the moon as a missile went unlearned. Things continued to go wrong. Here are but three of many:

1980—In the early morning hours of June 3, seventy-six B-52 nuclear bombers prepared to take off in response to warnings on display screens at a Strategic Air Command post in Omaha, Nebraska. Two missiles launched from Russian submarines in the North Atlantic were believed to be heading toward the United States. In this particular instance, a quick-thinking commander was able to determine that the threat was not real, so the counterattack was canceled. A later investigation found that the potentially disastrous false alarm had been caused by a malfunctioning forty-six-cent computer chip.

1995—On January 25, the Russian early warning system mistook an arctic research rocket for an incoming American missile. Military commanders advocated an immediate nuclear counterstrike against the United States. Fortunately, Boris Yeltsin, the Russian president, was able to overrule the military men, and we are still here to tell the story.

2018—On January 13, residents of Hawaii were thrown into panic by an alert in all capital letters on their television and smartphone screens:

BALLISTIC MISSILE THREAT INBOUND TO HAWAII.

SEEK IMMEDIATE SHELTER.

THIS IS NOT A DRILL.

It was a false alarm, later attributed to a poorly designed dropdown menu on the warning system's computer. But although the alert was broadcast in error, Hawaiians had been primed to anticipate a nuclear attack from North Korea. The significance of the incident was summarized in a news headline: "Hawaii False Alarm Hints at Thin Line Between Mishap and Nuclear War."[36]

"Star Wars": Reagan's Strategic Defense Initiative

Missile defense began to extend into space as early as 1960 with the first rudimentary attempt at using Earth-orbiting satellites to detect Soviet attacks. Land-based early warning systems could see only as far as the horizon, but eyes in the sky would be able to peer deep into the Soviet Union, all the way to the ICBM launchpads in Kazakhstan.

The first decade of these space-based efforts was largely a chronicle of failure. Not until the early 1970s did the technology begin to bear significant fruit. From that point forward, missile defense became ever more reliant upon infrared-detecting reconnaissance satellites in geosynchronous orbit.

In 1983, Ronald Reagan announced a bold new technological undertaking to create a "shield in space" that not only would warn against incoming missiles but would intercept and destroy them as well. The official name of the program was SDI—Strategic Defense Initiative—but it quickly became universally known as Star Wars. Star Wars' promise—that it could give total protection against nuclear attack—required the creation of space-based, robotic "kill vehicles" armed with nuclear death rays. It was undeniably a fanciful notion, but critics who pointed out its implausibility were reminded that numerous DARPA projects, though initially ridiculed as impossible, eventually achieved success. As the Gershwin song goes, "They all laughed at Christopher Columbus..."

Nonetheless, when Reagan announced the birth of SDI, the technology was barely in the conceptual stage. Thousands of skeptical physicists sharply criticized it, and less than ten years later, in 1992, the original core of its technology was scrapped. By that time, however, the Soviet Union had ceased to exist, and numerous pundits credited Stars Wars as the primary factor in the USSR's downfall. The fear that SDI *might* succeed, they said, bankrupted the Soviet economy as it struggled to keep pace in the arms race. The evidence supporting that superficial analysis is thin.[37] But even on its own terms, the triumph it claimed would have been political, not scientific or military.

Edward Teller and the Death Ray

Although Star Wars science turned out to be of little value, its great expense and symbolic weight make it worth examining. It all began with Edward Teller, the patron saint of Cold War science and the military-industrial complex's most aggressive ally within the scientific community. Teller envisioned a hypothetical weapon he called Excalibur—a hydra-headed death ray that could blast any and all incoming enemy missiles out of the sky. The far-fetched proposal gained the energetic support of the right-wing Heritage Foundation, the think tank that set the agenda of the Reagan administration.[38] Reagan himself became Teller's most zealous evangelist.

The death ray concept was a decades-old science fiction staple that inspired a great deal of research into directed-energy weapons. The goal was to be able to obliterate distant targets with microwaves, lasers, particle beams, or . . . whatever. Teller's Excalibur featured X-ray lasers. Lasers, such as the familiar light-beam pointers used by lecturers and snipers, emit energy in the form of coherent electromagnetic radiation. Coherent waves' "peaks and valleys" are in lock-step alignment, as opposed to their randomness in most natural states. Because the waves do not diverge, their energy can be concentrated on a small spot at a great distance.

The spectrum of electromagnetic radiation includes radio waves, visible light, and X-rays. The lecturers' pointers utilize the visible-light portion of the spectrum. The beams they produce are not energetic enough to kill humans or destroy sizable objects, but you would not want to point them into your eyes. Laser eye surgery uses ultraviolet waves, which are more potent than visible light but less so than X-rays.

Because X-rays pack much more energy than visible light, an X-ray laser could conceivably function as a death ray. Teller's Excalibur was to incorporate dozens of X-ray lasers wrapped tightly around a thermonuclear bomb. Exploding the bomb would generate the X-rays that would then be focused by the lasers, which could autonomously aim themselves at incoming missiles. Just one of these weapons, Teller exclaimed, "could generate as many as 100,000 aimable beams" that could shoot down "the entire Soviet land-based missile force."[39]

The scheme would require aiming a hundred thousand bursts of X-rays emitted from a thermonuclear explosion with one-millionth of a degree of accuracy. Small wonder that thousands of physicists pronounced it "not technically feasible."[40] Teller dismissed the criticism, asserting that it was simply a straightforward engineering problem. To understand the mind that produced this phantasmagoria of horror, a clue is offered by the subtitle of a Teller biography: *The Real Dr. Strangelove*, comparing him to the archetypical mad scientist willing to blow up the world in service of a misanthropic ideological agenda.[41] Teller's ideological allies in the Heritage Foundation organized a Coalition for SDI that boasted two hundred member organizations and claimed to represent the majority view of the scientific community.

It should not be surprising that many physicists beholden to the military-industrial complex for their livelihoods might support—or at least not publicly oppose—the pro-SDI point of view. Far more impressive was the unprecedented protest of scientists on the opposing side of the controversy. Some seven thousand scientists signed a pledge not to accept money for SDI research, which they condemned as "deeply misguided and dangerous." Those who made that vow, the *Washington Post* reported, represented 110 research institutions and included "57 percent of the combined faculties of 20 of the nation's highest-rated university physics departments." Among them were fifteen Nobel laureates in physics and chemistry.[42]

The protest of independent scientists, however, was no match for the massive flow of government funds that attracted thousands of university researchers willing to work on the program. This particularly grotesque distortion of American science flourished because the Heritage Foundation prevailed against the considered advice of independent physicists. The success of missile defense schemes in generating billions of dollars in guaranteed profits for weapons manufacturers ensured their survival—in one form or another.

The Shape-Shifting Ghost of Star Wars in the Post–Cold War Era

The collapse of the Soviet Union in the early 1990s removed the primary rationale for spending hundreds of billions of dollars on missile defense systems. But although the USSR had disappeared, the economic imperative driving war spending had not. The military-industrial complex would generate new justifications for continuing the high-priced projects in new guises.

It was a thoroughly bipartisan venture. After Reagan and the first President Bush, the Clinton administration changed the program's name to the Ballistic Missile Defense Organization, but continued to fund it. Bush the Younger continued down the path toward the militarization of space; his administration's National Missile Defense program was aptly nicknamed "Son of Star Wars."

In 2018 the *New York Times*, which often offers a rational corrective to the shortsighted rapacity of some sectors of the American corporate oligarchy, warned against "the dangerous illusion of missile defense." Its editorial board declared, "After more than 30 years of research and more than $200 billion," the missile defense program continues to be "riddled with flaws."[43]

And then Trump stepped up to the plate with Grandson of Star Wars. On January 17, 2019, the Pentagon's useful idiot gave a speech that "embraced Reagan's Star Wars plan of putting weapons in space to shoot down enemy missiles," and called for spending billions of dollars on it.[44]

The underlying premise of the antimissile programs—that they could "hit a bullet with a bullet"—was from the beginning, and still is today, fraudulent. Despite the hundreds of billions of dollars spent on it, the technology to accomplish its promises never existed. Nonetheless, at the end of the second decade of the twenty-first century, the precious resources of American science were still being wasted on research pursuing the Star Wars fantasies of administrations past and present.

CHAPTER 15

Video-Game War

THE DRONE, OR UNMANNED AERIAL VEHICLE (UAV), has brought about an unprecedented revolution in the way war is waged. Instead of soldiers risking death on a battlefield, high-tech warriors can now vanquish foes by wielding remote-control weapons from thousands of miles away. The military's preferred euphemism for this impersonalized mass murder is "offset technology."

It has also been called video-game war: a console operator in Nebraska targets humanoid avatars on a computer screen, squeezes a trigger on a joystick, and instantly rains death upon dozens of flesh-and-blood human beings in Pakistan. The act of killing is sanitized and dehumanized by geographical and emotional distance. Traditional military values nostalgically exalted by old soldiers, such as courage and honor, lose their meaning in the context of drone warfare. Even the illusion of glory is gone.

Predators, Reapers, and Other Purveyors of Doom
American civilians tend to think of drones as relatively small contraptions, like the cute little quadcopters we sometimes see buzzing overhead at sporting events and elsewhere. But the military drones' wingspans often match those of midsize jet aircraft, and they can launch cruise missiles and drop 500-pound bombs. The air force's Reaper drone, for example, is 36 feet long and has a 66-foot wingspan. It can carry 3,750 pounds of attack munitions, including laser-guided air-to-ground Hellfire missiles, the weapon most frequently launched from US drones.

The names the Pentagon gives its drones and their accoutrements are not subtle. *Predator. Reaper. Hellfire.* No effort is made to sugarcoat their purely malevolent functions. A Predator drone pilot boasted in an interview that the Hellfire missiles redesigned for use with the Predator had been upgraded to include "an even nastier antipersonnel bonus" than earlier models:

> When the two charges, wrapped in a sleeve of scored steel detonated, the sleeve shattered along its scored lines and blasted out razor-sharp shrapnel in all directions to slice and dice anyone within a twenty-foot radius (depending on the surface). Even those out to fifty feet might not escape its wrath.[1]

The Predators, however, proved insufficiently deadly for Pentagon purposes, which led to the introduction of the Reaper drones in 2007. Reapers can fly twice as fast, twice as high, and nine times farther than the Predators, and can fire twice as many missiles, making them far more effective hunter-killers. In March 2018 the air force officially announced the retirement of the Predators and full transition to the Reapers.

When Is a Weapon Not a Weapon?

The Reaper, the air force assures us, is not a mere weapon; it is a *weapons system*. Keeping dozens of pilotless aircraft in the air simultaneously, twenty-four hours a day, requires coordination and communications technology with a level of sophistication that the words "remote control" only begin to hint at.

At a typical drone base in the Middle East, Reapers operate in groups of four. The ground controllers depend upon satellite links for real-time video feeds and other data, seamlessly merged from the quartet of Reapers. The drones' infrared-capable cameras allow the operators, and their commanders in the field, to survey everything, day and night, going on over a wide swath of territory. In a four-Reaper squadron, each one requires two operators on the ground—one to pilot the UAV and one to manage its data collection and dissemination. The two-person teams may be at a nearby air base, or they could be thousands of miles away, at consoles in the United States. An overall commander would most likely be overseeing the operation from a third remote location. Furthermore, the two-person teams routinely switch off with other teams in midflight. One duo may handle takeoff and landing while another conducts the hunter-killer part of the mission.

The many billions of dollars the Pentagon has devoted to drone technology are small potatoes in the context of total military expenditures, but the price is high in absolute terms and rising rapidly. The DoD planned to spend $6.97 billion on UAV research in 2018, a 21 percent increase over 2017.[2]

A Brief History of Military Drones

While drones may seem to have come out of nowhere in recent years, in fact their development as surveillance tools and offensive weapons had secretly been in the works for many decades. Drones only came to the attention of the general public after the turn of the twenty-first century, but UAV technology was already a significant focus of military research in the early-to-mid-1960s.

That research, conducted in Vietnam in the heat of war by DARPA, was carried out under the strictest possible conditions of secrecy. When drones in large numbers were deployed in the field, following the 1964 Bay of Tonkin incident, they could no longer be hidden from outsiders' eyes. More than three thousand drone missions were flown during the Vietnam War, and many were shot down. The Pentagon belatedly acknowledged the existence of their UAVs in 1973.

The drones in Vietnam were not offensive weapons. They could not carry guns, bombs, or rockets. They were used to conduct aerial reconnaissance, to drop propaganda leaflets, and to act as decoys to confuse North Vietnamese surface-to-air missiles. The earliest versions were relatively primitive. Over time, researchers were able to perfect their three-dimensional control capabilities and provide them with the ability to transmit real-time surveillance video. But the major advances toward the creation of Predators and Reapers only became possible in the 1980s, as computers became exponentially more powerful and more miniaturized.

The advanced technology necessary to arm drones with missiles required several decades to achieve. The air force only began to pursue that goal in the late 1990s. Armed drones were first used in battle in Afghanistan in late 2001 and then rapidly became ubiquitous. The Bush-Cheney administration extended their use beyond declared war zones, a monumental step in the progression toward permanent, globalized American warfare.

Obama's "Drones for Peace"

A total of fifty-seven drone strikes were authorized on Bush-Cheney's watch, but that was just for openers. Barack Obama in his first year in office carried out more drone strikes than George W. Bush did in his entire eight-year tenure. Obama earned the ironic sobriquet "President Drone" by launching almost ten times as many strikes—563—as Bush and Cheney had.

Defenders of Obama's image as a man of peace cite the statistic that when he took office there were two hundred thousand US troops in Iraq and Afghanistan, and when he left there were only about fourteen thousand in the two countries. That was due, however, not to a decline in war making but to a transformation in the way the United States wanted to wage war. In May 2013, in the first speech of his second term, Obama quoted James Madison—"No nation could preserve its freedom in the midst of continual war"—and pledged to end the War on Terror. To accomplish that, he would step up the use of drones. The militarization of American science facilitated the replacement of combat troops ("boots on the ground") with increasingly high-tech weaponry, including drones.

Obama also vastly expanded the presidential prerogative to wage war with no congressional oversight, a precedent the Trump administration has warmly embraced. In October 2017 more than 240,000 troops in at least 172 countries and territories were fighting what the *New York Times* calls "America's Forever Wars." American forces were "actively engaged" not only in Afghanistan, Iraq, Yemen, and Syria, but also in Niger, Somalia, Jordan, Thailand, and elsewhere. "An additional 37,813 troops serve on presumably secret assignment in places listed simply as 'unknown.' The Pentagon provided no further explanation."[3]

The construction of new drone bases has been a major feature of this burgeoning global American military presence. Although for the most part hidden from the American public, one such base has received some news coverage. As of April 2018, Air Base 201—comprising 2,200 acres, including a 6,800-foot-long runway and 39 acres of paved airfield—was "rising from a barren stretch of African scrubland" in a remote area of Niger.[4] Its purpose is to accommodate drone operations from Senegal to Sudan.

Assassination by Remote Control

In February 2002 the CIA attempted its first "targeted killing" (the official euphemism for assassination by drone), but it didn't go as planned. It took place at a site called Zhawar Kili, in rural Afghanistan, and the intended target was Osama bin Laden. A Predator drone fired a Hellfire missile at a group of people including "a tall man" the CIA thought was bin Laden. It wasn't.

Journalists later identified the tall man as a local villager named Daraz Khan. He and the two other innocent victims killed in the strike, Jehangir Khan and Mir Ahmed, were scrap-metal scavengers, eking out a meager living by selling bits of bomb and missile detritus left by previous US air strikes. Military officials, although forced to admit that three men whom they had not intended to kill were dead, persisted in describing the victims as "appropriate" and "legitimate" targets.[5] The Zhawar Kili incident stands as an almost perfect exemplar of the ongoing era of remote-control assassination, with its wholesale slaughter of foreigners and its lack of accountability. The only anomaly was that the dead civilians could be identified by name.

Obama created another ominous precedent by extending the practice of extrajudicial assassination to include American citizens. The first victim was a Muslim cleric named Anwar al-Awlaki. Awlaki had been born, raised, and educated in the United States. Following the terrorist attacks of September 11, 2001, he was a prominent spokesperson for moderate Muslims who distanced themselves from radical Islamists. Later, however, he became radicalized and attracted a large internet following for his fiery denunciations of the United States as an implacable enemy of Islam. He left the United States in 2002, and by 2004 was operating a web-based ministry in Yemen.

Obama ordered the assassination of Awlaki in April 2010, and it was carried out by a CIA-led drone strike (which also killed another American, Samir Khan) on September 30, 2010. Two weeks later the CIA killed Awlaki's sixteen-year-old son, Abdulrahman al-Awlaki, also a US citizen.

The Trump administration's first major war crime was a January 29, 2017, raid in Yemen that killed thirty people, including Awlaki's eight-year-old daughter, Nawar al-Awlaki, who was also an American citizen.

"Thou Shalt Not Kill"?

Ever since the concept of legality first made its appearance in human affairs, civilized societies have placed extrajudicial assassination outside the law. That only changed in the twenty-first century, when the Bush-Cheney administration explicitly legalized "targeted killings" as a tactic in its War on Terror.

Earlier, in the 1970s, in the wake of the Watergate scandal, the American public was shocked to learn that the CIA had been routinely conducting assassination operations against foreign leaders since 1945. The spy agency's greatest hit (so to speak) was the 1961 assassination of Patrice Lumumba, the hero of Congolese independence and his country's first prime minister. The revelations led to congressional hearings that resulted in an executive order issued in 1976 by president Gerald Ford, declaring: "No employee of the United States government shall engage in, or conspire to engage in, political assassination." The intention was not to proclaim a new principle but to codify a legal norm long taken for granted.

In the wake of the September 11, 2001, attacks, however, the Bush-Cheney administration was able to parlay the public's fear of terrorism into an explicit rejection of the ban on political assassinations, opening the way to the era of targeted killings by drones. It was not until Barack Obama took office, however, that the assassination program was officially acknowledged.

On the third day of 2020, the Trump administration took another giant step in normalizing targeted assassinations by ordering the killing of a top-ranking official of a sovereign country, Iran, by means of a drone strike in another sovereign country, Iraq. The execution was accomplished by a Reaper; its target was Major General Qassim Soleimani of Iran. A leading Iraqi military chief, Abu Mahdi al-Muhandis, and at least three other people were also killed. This was a cold-blooded war crime that Trump not merely acknowledged but bragged about.[6]

How the Targets of Assassination Are Chosen:
The Disposition Matrix

There are two basic categories: "signature strikes" and "personality strikes." A personality strike is one in which specific individuals, identified as dangerous terrorists, are named and hunted down. A signature strike is much more general and impersonal. All military-age males in a foreign strike zone are designated as enemy combatants and are thereby deemed fair game for assassination. Whether condemned to death as terrorists or combatants, none of the targets are granted any semblance of due process before being executed.

With regard to personality strikes, in which individuals are targeted by name, the *New York Times* reported in May 2012 that President Obama "placed himself at the helm of a top-secret 'nominations' process to designate terrorists for kill or capture, of which the capture part has become largely theoretical."[7] Obama's roll call of doom, officially named the "disposition matrix," was casually referred to as his "kill list." Candidates for assassination were proposed to him by some two dozen national security officials at weekly meetings called "Terror Tuesdays." Because the meetings were conducted in secret, not much more is known about them, including the names and number of names on the kill list at any given time.

The American military's claims to carefully target only "militants" are belied by its notorious double-tap tactic. After blowing up homes where alleged militants gathered, drone operators would wait until neighbors came to the victims' aid and then dispatch a second Hellfire missile to kill the rescuers. The murder of misidentified individuals or innocent bystanders is especially outrageous, but assassinating *anybody for any reason*—be they "militant" or not—is no less unlawful. As investigative journalist Rebecca Gordon reminds us, "Executing foreign nationals without trial in other countries is itself wrong and illegal under U.S. law, as well as that of other countries where some of the attacks have taken place, and of course, international law."[8]

All of the twenty-first century US administrations have flouted the law by appeal to the 2001 congressional act entitled Authorization for the Use of Military Force (AUMF). Despite its explicitly narrow application to "nations, organizations, or persons" who "planned, authorized, committed, or aided the terrorist attacks that occurred on September 11, 2001," the AUMF has been indiscriminately stretched to justify innumerable drone assassinations of people suspected on the skimpiest of evidence of connections to terrorist activities.

Just before leaving office, Obama issued executive orders designed to limit his successor's ability to order drone assassinations, but Trump's lawyers swiftly overrode them with executive orders of their own. Drone strikes then dramatically increased. Accurate numbers are difficult to come by, but one comparison of drone strikes in Yemen, Somalia, and Pakistan during the two administrations' first two years in office revealed this score: Obama 186, Trump 238.[9] A *New York Times* editorial on American drones also observed that "under President Trump, airstrikes have surged in Afghanistan, Iraq and Syria."[10] And then in March 2019, Trump ramped up the secrecy surrounding the subject by rescinding a legal requirement that the government publicly report drone strike data.[11]

Biomimetic Microdrones and Other Technological Delights

As formidable as the Reapers are, new nanotechnologies are creating even more intimidating remote-control weapons at the other end of the size spectrum. Tiny insect-mimicking drones that operate in swarms, sneak into the private dwelling spaces of targeted victims, and blow their heads off with microexplosive bombs are rapidly transitioning from science fiction to science fact.

In 2013 the air force released an animated video extolling the lethal potential of biologically inspired Micro Air Vehicles (MAVs).[12] Whether such weapons are already operational, and how close they are to deployment, is unknown due to the secrecy enveloping the highly classified program.

"Flybots" are not the only miniaturized biomimetic drones being developed. The rapidly advancing field of nanobiotechnology is also creating diminutive mechanical spies and assassins with a full range of locomotive options. As a journalist who has investigated DARPA's current research plans commented, "The Pentagon's drones will fly, swim, crawl, walk, run, and swarm as they conduct missions around the globe."[13]

Blowback Potential

Shortly after their first appearance on the battlefield, American drones fell into the hands of their intended targets. It was then a relatively simple task to "reinvent" the drones by reverse engineering. Today, dozens of nations large and small, from China to Kazakhstan and from Russia to Myanmar, have weaponized drones in their military arsenals.

The rapid global proliferation of drone science and technology has increased the threat of blowback attacks. If American policymakers fear "Islamic terrorists"—or future shooting wars with China, Russia, or any other opponent—the problem is largely of their own making. They handed those adversaries a powerful death-dealing technology free of charge. The billions of dollars spent on UAV science and technology have not made Americans safer and more secure but have had the opposite result.

The threat of terrorism has, from the beginning, been largely a mirage, grossly exaggerated by US policymakers and a compliant mass media to instill fear in the American public. Occasional highly publicized terrorist incidents feed the illusion that the Pentagon and CIA are defending us from a dangerous enemy, and need every penny of the bloated war budget to do so. But a military challenge from a more formidable opponent would be another matter altogether. Meanwhile, the United States is no closer to concluding its "forever wars" than it was in 2003, when George W. Bush posed triumphantly on the deck of an aircraft carrier under a banner that declared "Mission accomplished!"

And while drone technology and the proliferation of remote-control warfare have already created a nightmare scenario for humanity, there is much worse on the horizon. If there is a class of weapons more to be dreaded than remote-controlled drones, it is drones that are *not* remotely controlled.

CHAPTER 16

Lethal Autonomy

BEFORE WE'VE ABSORBED THE SHOCK of the drone revolution, another menacing military technology is upon us. The coupling of remote-strike capability with artificial intelligence throws open yet another Pandora's box for humanity. Remote-control murder was bad enough—now we are faced with fully automated murder.[1]

Allusions to the *Terminator* films are impossible to avoid as we face the prospect of robotized war fighters. In real life, the phrase used by military policymakers to describe the goal of these machines is "lethal autonomy." Astute observers could see this coming as long ago as 2011. A *Washington Post* article envisioned "the future of the American way of war" as one in which "drones hunt, identify and kill the enemy based on calculations made by software, not decisions made by humans."[2] In 2012 the Defense Science Board, a highly influential advisory group within the Department of Defense, issued a report describing how "unmanned systems" were already "making significant contributions to DoD operations worldwide," and urging the Pentagon to "use autonomy more aggressively in military missions."[3]

By late 2016 the march toward lethal autonomy had apparently passed the point of no return. "The Pentagon," the *New York Times* reported,

> has put artificial intelligence at the center of its strategy to maintain the United States' position as the world's dominant military power. It is spending billions of dollars to develop what it calls autonomous and semiautonomous weapons and to build an arsenal stocked with the kind of weaponry that until now has existed only in Hollywood movies and science fiction.[4]

Although almost everything having to do with autonomous weapons research is highly classified, the American military has openly declared its intention to devote major attention and resources to it for years to come.[5]

AI and the Military

The technology of lethal autonomy is closely linked to the field of artificial intelligence. AI research today is based on collaboration between computer science and neuroscience—an often rocky and incompatible relationship, with expertise in machine technology on one side and the workings of the human brain on the other.

An aspect of AI of particular interest to the military is facial recognition, a key step in enabling robot drones to identify targeted individuals from photographs of their faces, pick them out of crowds, follow them home, and kill them—autonomously, without the guidance of human operators. Although it may not have occurred to most civilians, the ability of their computer or phone to put names to faces in their photographs is a product of military research with which they have unwittingly collaborated.[6]

American civilians may also serve as guinea pigs in facial-recognition research in less voluntary ways, as a *Wall Street Journal* headline announced: "Artificial Intelligence Could Soon Enhance Real-Time Police Surveillance."[7] In the near future a cop may pull you over, point his body camera at your face, address you by name, and know a great deal about you.

Plans to have facial-recognition body cams in the hands of local police departments by the fall of 2018 encountered public resistance that forced a retreat, at least for the time being. In June 2019, California legislators proclaimed they might impose a statewide ban on the devices. Meanwhile, the major manufacturer of police body cameras declared that it would not provide facial-recognition capability in its products.[8]

Facial-recognition body cams originated in a DARPA-funded program named SyNAPSE.[9] This venture into "electronic neuromorphic machine technology" is aimed at transcending today's AI to achieve AGI—artificial *general* intelligence—by creating a computer modeled on the mammalian brain, emulating ten billion neurons and a hundred trillion synapses. How far along DARPA and its contractors are toward that lofty goal is a closely guarded secret. As a discerning investigative journalist has observed, technologies of this sort are "unveiled to the public only after they have created a revolution in military affairs."[10]

Meanwhile, Down in Silicon Valley . . .

Not all of the advances in artificial intelligence have come directly from the military, however:

> The defense industry no longer drives research the way it did during the Cold War, and the Pentagon does not have a monopoly on the cutting-edge machine-learning technologies coming from start-ups in Silicon Valley, and in Europe and Asia.[11]

That observation is superficially valid but also misleading. Corporations wanting to lower production costs by replacing human labor with robots do indeed represent a strong civilian motive to pursue AI research. But Silicon Valley's contributions were predicated on earlier military research. If, for example, you thought that Steve Jobs or computer scientists in his employ gave Siri the ability to engage in witty repartee with humans, you would have been wrong. Siri was a product of a DARPA program called PAL, "Personalized Assistant that Learns," the largest AI research project in American history. In DARPA's own words, PAL's mission was "to make military decision-making more efficient and more effective." It also, however, "led to the 2007 launch of Siri Inc., later acquired by Apple Inc., which further advanced and then integrated the Siri/PAL technology into the Apple mobile operating system."[12]

Furthermore, the extent to which the Pentagon "drives" the research process is irrelevant, because the technology serves military as well as civilian agendas. DARPA does not have to control all AI research to advance the military quest for lethal autonomy. That is illustrated by Silicon Valley's prominent role in self-driving-vehicle research. The most familiar names in the Big Tech universe—Google, Apple, Amazon, Tesla, and Uber—have invested heavily in the effort to consign human drivers to the dustbin of history.

Cars and Trucks without Drivers

When did the serious prospect of driverless vehicles first impinge upon your consciousness? Unless you have a special interest in robotics, probably not before 2015. Now, however, as a report in *Wired* magazine declares:

> self-driving cars aren't just here. They are, it seems safe to say, just about everywhere—roaming the streets of San Francisco, New York City, Phoenix, Boston, Singapore, Paris, London, Munich, and Beijing.[13]

Google quietly began research into self-driving automobiles in 2009. By 2015 it was loudly trumpeting the results, claiming that its autonomous vehicles had logged way more than a million miles, mostly on big-city streets. (The "way more" became Waymo, the name of the Google subsidiary in charge of the research.) The first self-driving truck also hit the streets in 2015. Little more than a year later, an Uber subsidiary announced that one of its fully automated vehicles had traveled 240 miles along Interstate 25 in Colorado to deliver a truckload of beer. In early 2018 yet another milestone was reached: the first coast-to-coast driverless truck journey—a 2,400-mile passage from Los Angeles to Jacksonville, Florida. Shivers no doubt went down the spines of three and a half million American truck drivers as they contemplated the disappearance of their livelihood.

Meanwhile, military AI research had preceded these Silicon Valley triumphs by decades. DARPA funded a Strategic Computing Initiative from 1983 to 1993, which, while not producing the high level of machine intelligence it aimed at, was by no means a total failure. It created the scientific basis for successful self-driving technologies that the Silicon Valley entrepreneurs later drew upon.

DARPA's later role in encouraging civilian collaboration in robotic research was uncharacteristically public. The agency held well-publicized contests—the DARPA Grand Challenges—with significant cash prizes. The challenges were set as races between robotic devices submitted by teams. The first race, in 2004, ended with no entrant reaching the finish line, so no prize was awarded. Subsequent DARPA challenges, however, were more fruitful. There was a direct line between the second challenge, in 2005, and the 2009 appearance of Google's pioneering self-driving

car. Sebastian Thrun, who led the team that won the 2005 DARPA challenge, subsequently became head of the Google project.

By 2007, the competitions had attracted many major university
and corporate participants. The prizes were $2 million for the winner,
$1 million for the runner-up, and $500,000 for third place. But more
notably, DARPA also gave eleven teams *preliminary* funding of $1 mil-
lion each. The 2007 winner was a team of researchers and engineers
from Carnegie Mellon University and General Motors. Later challenges, which continue to the present, have emphasized increasingly sophisticated automated tasks. The announcement for the 2019 DARPA
competition solicited proposals for the "development of micro-to-milli
(insect scale) robotic technologies."[14]

The DARPA challenges succeeded in drawing leading engineering
schools and corporate researchers into the development of advanced
autonomous weaponry. Furthermore, the public nature of the competitions from 2004 to the present have also served an important PR purpose: by publicly sponsoring the development of self-driving cars and
trucks, and especially robotized emergency-response vehicles, DARPA
hoped to soften its image as a purely secretive agency interested only in
the technology of death.

Prosthetic Hype versus Reality

DARPA likewise sought humanitarian cred for creating miraculous prosthetic limbs for amputees. You may have seen the story on *60 Minutes:*
a spectacular, emotional account of American soldiers whose missing
arms were to be replaced, thanks to DARPA, by computer-controlled
myoelectric limbs with fully functioning hands and fingers. The director
of the research effort compared it to the Manhattan Project in scale, and
said it involved "well over 300 scientists; that is, engineers, neuroscien-
tists, psychologists."[15]

The *60 Minutes* segment was impressive theater. But when investigative journalist Annie Jacobsen pursued the story behind the story, she
discovered that

> when the cameras go off, the arms usually go back to the DARPA
> laboratories, where they generally sit on shelves. "Most of us strap

back on our Captain Hook arms," said one participant, who lost
an arm in Iraq.[16]

The wounded veterans were window dressing to divert the public
eye from advanced lethal autonomy research. DARPA's ulterior motive
was to develop humanoid limbs for war-fighting robots.

Nonetheless, DARPA science had undeniably produced a technological marvel of immense potential value to many suffering amputees. In
2014 the FDA approved its production for the commercial market. And
then the market did what the market does. It put a huge price on the artificial limb that places it out of financial reach for many. The "average lifetime
cost for prosthetics and medical care for loss of a single arm," according to
a US Department of Veterans Affairs estimate, would be $823,299.[17]

And there you have the tragedy of American science in microcosm:
great potential to improve human lives, thwarted by militarization and
market forces.

From Biomimetics to Cyborgs

DARPA projects to invent machines that mimicked spiders, flying insects, birds, fish, dogs, cheetahs, mules, and humans morphed smoothly
into the creation of biological-mechanical hybrids. Surgically implanted
pacemakers have been around since 1958, but advances in nanotechnol-
ogy at the turn of the twenty-first century brought forth a new era. The
ability to insert miniscule machines into living organisms inaugurated
the age of biohybrids, another name for which is a sci-fi staple: *cyborgs.*

Implanting electrodes in the wings of moths to control their flight
paths may seem like nothing more than a weirdly esoteric pursuit.
DARPA scientists accomplished that in 2014, in typically spectacular
fashion. They inserted electrodes during the insects' pupa stage; by the
time the full-fledged moths emerged, the electrodes were fully integrated
into their little bodies. They were cyborg moths.

The Pentagon wasn't paying DARPA to fool around with remote-
controlled insects out of idle scientific curiosity. The motive underlying
this research was revealed several years earlier in an insight by RAND Corporation analyst Benjamin Lambeth. In a 1996 paper, Lambeth suggested
that biomimetic minidrones would make splendid assassins. All it would

take would be to equip them with "micro-explosive bombs" that could "kill moving targets with just grams of explosive."[18] In the pursuit of lethal robots, murder-moths were just the beginning.

Exoskeletons for Supersoldiers

How would one go about engineering a supersoldier? In 2001, DARPA began R&D efforts to produce what it called an "exoskeleton" to enhance the physical abilities of ordinary GIs. Unlike medieval military exoskeletons—suits of armor—the modern variant would not weigh down the wearer but would put spring in his or her step.

In 2014 a *Forbes* reporter was invited to try on a prototype of "a low power (under 100 watts), lightweight (40 lbs), under-the-clothing exoskeleton that lets soldiers walk, run or climb farther and faster without extra effort." The apparatus was controlled by a computer attached to a backpack and included motorized leg braces tucked into military boots, which the journalist described:

> The computer reads leg movements and kicks in the right hydraulic boost (with the reassuringly RoboCop-like *zzzt-zzzt-zzzt* sound) to kick my legs forward just enough so I don't have to work as hard. Walking quickly in the suit is almost like being buffeted by a light wind at your back. When I break into a light jog, the motors get my knees up just a bit higher than I normally would.[19]

The exoskeleton also doubles the number of bench presses and push-ups its wearer can do. Although it may seem to presage a dystopian future policed by RoboCops and Terminators, the exoskeleton represented a considerably more benign innovation than some of DARPA's other projects. Completely external to the human body, it did not cross the line into transhumanism.

Transhumanism, Part I: Warriors on PEDs

One obvious way to try to create an *Übermensch* would be the pharmaceutical route that transforms athletes into jacked-up caricatures of their natural selves. DARPA did not neglect the great promise of battlefield performance-enhancing drugs. The agency's 2004 Peak Soldier Performance project was a "better warfighters through chemistry" research effort. Its purpose was to create stimulants that would give soldiers the ability to endure on the battlefield for five days without food. In pursuit of that goal, a British defense industry news service reported, "no stone or genome was left unturned."[20]

The research aimed to produce nutrition supplements to build endurance through lowering body temperature and enhancing the effectiveness of mitochondria, which convert sugars into chemical energy to power the body's cells. Biochemists strove to genetically alter mitochondria to have them

> feed off fat-based ketones, leading to much smaller, ketone-based rations being capable of not only keeping soldiers on their feet, but having them operate at their peak for days at a time. Initial laboratory tests demonstrated that rats given the treatment were able to run for extended periods of time.[21]

A 2003 report issued by the office of president George W. Bush revealed yet another chemical means of "human enhancement" under consideration. The report bore a title that would have made George Orwell either cringe or smile: "Biotechnology and the Pursuit of Happiness."[22] *Army Technology*, the British publication quoted above, analyzed the report and observed that it

> alluded to the potential development of drugs that could suppress the fear and inhibition of soldiers, effectively turning them into killing machines capable of acting [with] impunity. Given the clamour from agencies including the US Army, US Air Force and DARPA to instigate research into [PEDs] and the explicit use of pharmaceuticals capable of enhancing performance, it must be considered a matter of when, rather than if soldiers are issued with such drugs.[23]

Transhumanism, Part II: From GI Joes to GI-GMOs?

DARPA has also acknowledged interest in "development of techniques and tools to rapidly sequence, synthesize, and manipulate genetic material,"[24] raising the specter of a military-sponsored neo-eugenics, or "directed evolution" for the transgenic GMO era.

In a 2016 interview, deputy defense secretary Robert Work hinted that genetic modification techniques could be used to produce superior soldiers. When asked directly if such research is currently under consideration, Work reportedly "hedged a bit on whether the US military would genetically modify humans, saying 'that is really, really troubling.'"[25]

Do DARPA and the Pentagon harbor neo-eugenic ambitions? If so, how far along are they toward achieving them? Official secrecy makes it impossible to know. It is evident, however, that GI-GMOs are on their minds, and that they have ample funding and political backing to pursue such research with no public oversight.

Transhumanism, Part III: Microchips in Brains

DARPA has long been funding research involving implanting electronic microchips into human brains. That is not in dispute; the agency proudly and publicly acknowledges it.[26] A *Forbes* magazine headline announced: "DARPA Invests $18.3 Million in Brain Implant Startup That's Building 'A Modem for the Mind.'"[27] What is disputed is the motive underlying its many "brain-computer interface" programs. DARPA claims its intentions are purely altruistic: to heal traumatic brain injuries of wounded soldiers.

Critics, however, warn that DARPA is in fact aiming at the creation of cyborgs with superhuman strength, endurance, intelligence, and an array of powers to maximize their efficiency as killing machines. The superpowers might allow "enhanced human" soldiers to see in the dark, reduce or eliminate their need for sleep, and wirelessly receive and transmit electronic data.

The Quest for "High-Quality Brain Control"

One of DARPA's projects calls for a brain implant that can "record high-fidelity signals from 1 million neurons."[28] Previous researchers had recorded signals from a few hundred neurons, and used them in devices implanted in amputees' brains to help control prosthetic arms and hands.

DARPA intends to make the communication two-way. The embedded microchips would not only record signals from neurons but would also transmit computer-generated signals back to neurons. The stated aim is to create "neural prosthetics" that would provide visual information to blind people and auditory signals to the deaf. Who could object to such noble pursuits?

But those same neural prostheses could easily be repurposed to perform less humanitarian functions. In 2017 a defense industry news service reported that leading military scientists were researching ways to bypass "triggers, throttles, keyboards" and "plug the computer directly into the human brain." Instead of "the human operator *physically* touching a control," the implanted chip would "respond to the human's brain cells forming the *intention* to use a control."[29] Another potential application is to alter "moods" that affect soldiers' "morale" (translation: make them more gung-ho and aggressive). Yet another aims at "managing stress" in "cognitively challenging environments"[30] (translation: tranquilize soldiers' fear of risking their lives in battle and their inhibitions and remorse about killing).

Morality, Ethics, and the Terminator Conundrum

General Paul Selva, vice chairman of the Joint Chiefs of Staff during both the Obama and Trump administrations, calls concerns about the morality of cyborg creation "the Terminator conundrum," and asks, "When do we want to cross that line as humans? And who wants to cross it first? Those are really hard ethical questions."[31] But can the Pentagon's monitoring of its own morality and ethics be depended upon to deter it from using drugs, brain implants, and neo-eugenics to engineer young army recruits into supersoldiers?

The issue of cyborg ethics is subordinate to the more general problem of lethal autonomy. Should robots be entrusted with life-and-death power over human beings? If they commit war crimes (attacking schools or hospitals), who can be held to account? Should they be empowered to take actions that entail the risk of initiating a thermonuclear exchange? Two leading human rights organizations examined the issue and produced a report demanding that "governments should pre-emptively ban fully autonomous weapons because of the danger they pose to civilians in armed conflict." Robots, the report stated,

> would not be restrained by human emotions and the capacity for compassion, which can provide an important check on the killing of civilians. Emotionless robots could, therefore, serve as tools of repressive dictators seeking to crack down on their own people without fear their troops would turn on them.[32]

In July 2015 an open letter signed by thousands of AI and robotics researchers observed that

> autonomous weapons are ideal for tasks such as assassinations, destabilizing nations, subduing populations and selectively killing a particular ethnic group. . . . Just as most chemists and biologists have no interest in building chemical or biological weapons, most AI researchers have no interest in building AI weapons—and do not want others to tarnish their field by doing so.

It was imperative, the letter concluded, that an AI arms race be "prevented by a ban on offensive autonomous weapons beyond meaningful human control."[33]

This appeal fell on ears of the willfully deaf. The Pentagon's "context-based ethics" place the exigencies of military affairs above traditional moral concerns and excuse any and all criminality in the name of "national security." As for DARPA, it routinely hires bioethicists to critique its bioscience projects, but the agency is in no way obliged to heed the bioethicists' concerns or recommendations. Leaving that aside, *how ethical are the ethicists?* Do the bioethicists simply provide cover for morally indefensible research? And to what extent are they compromised by their materially rewarded involvement?

The bottom line in any discussion of the American military's ethicality and morality is the fundamental morality of its mission. Are its wars *just* or not? That will be considered in another chapter, which examines the doctrine of American Exceptionalism.[34] Meanwhile, no survey of science's contribution to the American way of war can be complete without considering the latest big thing: cyberwarfare.

CHAPTER 17

Is Cyberwarfare Really a Thing?

A PROVOCATIVE HEADLINE in a journal, *Breaking Defense*, devoted to "industry news, analysis, and commentary" asserts:

> "Cyberwar" Is Over Hyped:
> It Ain't War Til Someone Dies[1]

If cyberattacks don't kill people, should cyberweapons be included among the technologies of death? No one will ever die in cyberspace. That is a truism. But it is a dangerous illusion to think that cyberwar can be confined to cyberspace. A hacker hijacking a missile-armed drone could obviously cause a great deal of carnage in the physical world. And that only hints at the potential consequences of cyberwarfare. All military forces, national economies, power grids, and water supplies are now completely dependent on computers and computer networks. That means disrupting them could have violent, deadly repercussions.

One Face of Cyberwar: Sabotage

Among other things, cyberwarfare is a form of sabotage aimed at undermining "command and control," a key phrase in the military lexicon. Often abbreviated as "C2," maintaining command and control is the essential core of military doctrine. Without the ability to command and control troops and war matériel, a general is helpless.

Cyberattacks are particularly insidious because, as anyone who has ever depended upon a computer knows, unexpectedly losing command and control of it is frightening, frustrating, and disorienting. Military commanders experiencing a sudden loss of C2 in the heat of battle could be expected to react in erratic ways. Cyberwarfare, therefore, is an especially destabilizing practice. That poses a crucial question: Should one really want to make a thermonuclear-armed enemy *less stable?* Obviously not, which is why a rational society would rigorously prohibit any and all hacking into nuclear systems.

Unfortunately, the United States unwisely crossed that most perilous of lines well before the rise of Trumpian recklessness. Sometime around 2009, the American agency in charge of cyberwarfare, the NSA, joined with its Israeli counterparts to infect Iranian computers with a virus (the "Stuxnet worm") that sabotaged Iran's uranium-enriching centrifuges. Then in 2014, commander in chief Barack Obama personally ordered a clandestine cyberattack aimed at making North Korean nuclear rocket tests fail. In both cases, the disruptions were temporary, but they exacerbated international tensions. Fortunately, on neither occasion did the situation spin out of control.

The NSA is certainly aware that its cybersabotage cannot permanently stifle Iran's or North Korea's nuclear ambitions. Meanwhile, cyberattacks are notoriously prone to unintended consequences. What if, for example, a side effect of infecting North Korea's missile system had been a weakening of their safeguards against accidental launching? Nor can the seriousness of blowback potential be ignored. The American military enjoys a dominating advantage over the rest of the world in physical weaponry, but not in cyberwarfare, which is relatively inexpensive to wage. Even small countries can match the United States in hacking prowess, and the US is at least as vulnerable as its foes. As one American cybersecurity expert warned: "We've got glass infrastructure and we shouldn't be throwing stones."[2]

Throwing More Stones

Not surprisingly, the Trump administration has opted to throw more stones. In September 2018 ultrawarhawk John Bolton, Trump's national security advisor, announced a new policy that would loosen previous restraints on the ability of American military commanders to order cyberattacks.[3] The new White House directives eliminated restrictions requiring the departments of commerce, treasury, homeland security, and others to sign off on an offensive action before it can be launched. In particular, they further empowered General Paul Nakasone, NSA director, to initiate cyberoffensives.

This is all part of a "cyberdeterrence initiative" that echoes nuclear deterrence strategies of an earlier era. However, as David Sanger, a leading cyberwar correspondent, has observed,

> Mr. Bolton, whose concepts of deterrence were formed in the Cold War, is likely to discover what his predecessors learned: Almost every strategy that worked in deterring nuclear attacks does not fit the digital era, and even figuring out where an attack originated can be a challenge.[4]

The danger thus increases of an American retaliatory strike (for example) against China in response to an anonymous hack attack from anywhere. The ultimate consequences would be impossible to predict.

American Cybervulnerability

The Trump-Bolton announcement was remarkably ill timed. Less than a month later, a government agency issued a blistering report exposing the Pentagon's unreadiness for combat in the cyberspace arena.[5] American cyberdefenses, it revealed, were hopelessly vulnerable.

The report was based on a Government Accountability Office investigation. GAO-authorized hackers had been directed to test the defenses of eighty-six Pentagon weapons-development programs to determine how secure they were. The results were astonishing.

The hackers had little difficulty penetrating and seizing control of key weapons systems. The targets included DoD programs to create new missiles, submarines, fighter jets, and satellites. Among them were two of the three main delivery systems for thermonuclear weapons: the Columbia-class submarine[6] and the Ground Based Nuclear Deterrent.[7] Here are a few of the more egregious of the GAO report's findings, in its own words:[8]

> • Test teams were able to defeat weapon systems cybersecurity controls meant to keep adversaries from gaining unauthorized access to the systems. In one case, it took a two-person test team just one hour to gain initial access to a weapon system and one day to gain full control of the system they were testing.

- Once they gained initial access, test teams were often able to move throughout a system, escalating their privileges until they had taken full or partial control of a system. In one case, the test team took control of the operators' terminals. They could see, in real-time, what the operators were seeing on their screens and could manipulate the system. They were able to disrupt the system and observe how the operators responded.

- Multiple test teams reported that they were able to copy, change, or delete system data, including one team that downloaded 100 gigabytes, approximately 142 compact discs, of data.

- Poor password management was a common problem in the test reports we reviewed. One test report indicated that the test team was able to guess an administrator password in nine seconds.

- Multiple weapon systems used commercial or open source software, but did not change the default password when the software was installed, which allowed test teams to look up the password on the Internet and gain administrator privileges for that software.

- Multiple test teams reported using free, publicly available information or software downloaded from the Internet to avoid or defeat weapon system security controls.

- Test team activities were not detected at all during some assessments, including one case in which the test team operated for several weeks without being detected. One test report indicated that test team activities were not discovered even though the test team was deliberately "noisy" and was not trying to hide its activities.

- Another test report indicated that the intrusion detection system correctly identified test team activity, but did not improve users' awareness of test team activities because it was always "red." Warnings were so common that operators were desensitized to them.

In addition to cataloging problems they had discovered, the GAO analysts identified a reason why the problems may not be easy to solve. They noted that the Pentagon "struggles to hire and retain cybersecurity personnel" with sufficient expertise. Its officials complained that "once their staff members have gained experience in DoD, they tend to leave for the private sector, where they can command much higher salaries."[9]

The GAO investigation did not include nuclear weapons because they are under the aegis of the Energy Department rather than the Pentagon. One month earlier, however, another report, this time by a nongovernmental organization, expressed concern that the nuclear weapons programs are no less vulnerable to hostile hacker attacks.[10] Whether particular digital vulnerabilities are actually exploited or not, the real possibility that anonymous hackers could covertly commandeer the most lethal of weapons at any time ramps up the randomness and unpredictability of the ultimate threat to all of human existence.

The Other Face of Cyberwar: Espionage

Espionage is the primary element of cyberwar, even more essential than sabotage. Although they complement each other, their missions are not always fully compatible. Sabotage often aims at a quick act of destruction that exposes its existence and brings it to an end, whereas automated cyberspies can quietly lie low, hidden in dense thickets of code, for a long, long time.

The purpose of espionage is to covertly gather information about the activities and plans of an adversary. Over the past decade or so, unwelcome outsiders have increasingly infiltrated American financial and military computer networks. Who are the intruders and where are they from? They are legion, and their provenance cannot always be identified. Among the spectacular cases in which attribution is possible are North Korea's devastating 2014 hack of Sony's Hollywood division, and Russia's multifaceted hacking of the 2016 US elections.

The Center for Strategic and International Studies maintains a record of "significant cyber incidents." Its list for the first nine months of 2018 contained sixty-two items identifying mainly Russian, Chinese, North Korean, and Iranian culprits but also Pakistani, Lebanese, Turkish, Indian, and miscellaneous "state-sponsored" hackers of unknown origin.[11] This compilation was obviously shaped by an America-centric perspective; the mirror-image Russian or Chinese lists would showcase the works of American, British, and Israeli hackers.

The Cyberwar at Home

The passage of the USA Patriot Act in October 2001 opened the flood-gates to the unconstrained surveillance of the American population that National Security Agency whistleblower Edward Snowden revealed in 2013. The NSA conspired with private telephone companies and internet providers to listen in on the phone calls, record the metadata (phone numbers, time, and locations), and read the private emails of tens of millions of American citizens.

The NSA originated in 1952 as a branch of the Pentagon tasked with gathering *foreign* intelligence during the Cold War. However—as historian and cyberethicist Eben Moglen explains—after the dissolution of the Soviet Union, the "entire establishment of national security repurposed itself" to target terrorists instead of communists:

> We [the United States] no longer needed to spy upon an empire with 25,000 nuclear weapons pointed at us. Now we spied on the entire population of the world, in order to locate a few thousand people intent on various kinds of mass murder. Hence, we are told, spying on entire societies is the new normal.[12]

To accomplish this, Moglen adds, the spymasters first "corrupted the science" in ways that would "make their own stealing easier."

To spy on the entire population of the world required spying on the American population as well, but that had traditionally been ruled out as a fundamental violation of the US Constitution. Since 9/11, however, that principle has mainly been observed in the breach. Glenn Greenwald, the investigative journalist most responsible for making public the NSA documents leaked by Snowden, said they reveal that "the NSA's goal is to collect, monitor and store every telephone and internet communication that takes place inside the US and on the earth. It already collects *billions* of calls and emails *every single day*."[13]

All-pervasive NSA electronic eavesdropping has nullified Americans' right to privacy. Without the right to privacy, to anonymity, there can be no democratic self-governance. Former president Jimmy Carter gave expression to this verity when he declared, in the wake of the Snowden revelations, "America currently has no functioning democracy."[14]

The NSA has been able to continue unconstitutionally spying on Americans thanks to a secret body of law created by a system of covert surveillance courts. The 1978 Foreign Intelligence Surveillance Act (FISA) established a court system that would operate behind tightly closed doors, but limited its role to issuing warrants for surveillance of foreign spies. After 9/11, however, the powers and reach of the FISA courts grew exponentially. An investigative journalist described how the FISA court system "has quietly become almost a parallel Supreme Court." But "unlike the Supreme Court, the FISA court hears from only one side in the case—the government—and its findings are almost never made public."[15]

The Fourth Amendment to the US Constitution prohibits indiscriminate government spying on the American people, and the intelligence agencies swear they always operate within the law. The Snowden revelations exposed the enormity of that lie. The NSA was proven to have covertly conducted massive surveillance of the American public and had made secret treaties with intelligence agencies of other countries to share surveillance data with them. After the NSA was caught red-handed, the Obama administration described its crimes as "modest encroachments on privacy"[16] that had been approved by Congress and federal courts. That was blatantly false.[17] For its part, the NSA pledged to avoid future illegalities, but its assurances ring hollow.

Despite the 2015 USA Freedom Act purporting to restrict such activities, the NSA continued to gather and store vast amounts of data on Americans' phone calls and text messages. A 2018 report by the US director of national intelligence stated that in 2015 the NSA collected 151,230,968 "call detail records" from AT&T, Verizon, and other telecom companies. That number *more than tripled* in 2017, to 534,396,285.[18]

Data Mining

The technology most essential to NSA surveillance is data mining, an extension of the science of consumer targeting. It was not a DARPA product; it emerged from the pursuit of profit by Big Data corporations like Google, Facebook, Amazon, and YouTube. Data on the shopping habits of internet users can be harvested from website interactions and sold to marketers. Its value depends on its being connectable to identifiable individual consumers. In other words, profitable data mining *demands* eliminating anonymity on the internet.

The same data-mining techniques can be used to gather information on the political opinions of individuals. And of course, the NSA can use the same tools to keep a close eye on political dissidents. No one who uses a mobile phone or the internet in the United States today can have a realistic expectation of privacy. The NSA has the technical capacity to know who you are, where you are, where you've been, and who you've been communicating with.[19]

Quantum Computing—Cyberwarfare's Ultimate Weapon?

Meanwhile, another revolutionary technology looms on the horizon, threatening a major disruption of cyberspace if it transcends the conceptual stage and proves workable.

Quantum computing would replace the binary off/on switches ("bits") of the computers we use today with the ghostly flexibility of subatomic particles ("qubits") to create a new kind of supercomputer.[20] By shuffling qubits rather than bits, quantum computers are expected to be able to simultaneously process the same data in many different ways, making them millions of times faster and more powerful.[21] Consider this: a digital computer attempting to open a combination lock would try every possible number combination one at a time until the lock opened. The quantum computer could theoretically try every combination *simultaneously*.

The stakes in the race to create an operational quantum computer have been projected as not merely game-changing but game-*ending*: the winner would rule the world.[22] By overwhelming the encryption powers (that is, the "combination locks") of ordinary computers, the masters of the new technology could lay bare all of their enemies' secrets and thereby paralyze their defenses. Because all military resources are thoroughly computerized, a combatant with quantum computers would be able to command and control the nukes, the satellites, the communications—*everything*—of all foes lacking the new quantum technology.

Given that the quest for quantum computing will be colossally expensive, the only possible winners of this arms race will be major countries with large science establishments. The primary contestants are the United States and the People's Republic of China. China is already devoting billions of dollars to QIS—Quantum Information Science— and the US military is scrambling to catch up.[23]

As would be expected, American policymakers have begun beating the drums to terrify the public with the prospect that China might come out ahead in this competition. Weaponized Keynesianism is a primary driver of the campaign. An NSA spokesman declared that this threat "will require enormous investments by the United States and its allies to upgrade national security and surveillance systems."[24]

How real, and how imminent, is the danger represented by quantum computing? IBM, Google, and some tech start-up companies have produced small prototypes of qubit processors,[25] but whether fully operational quantum computers will ever become a reality remains a debatable issue. A report by the National Academies of Sciences offers some perspective:

> The field of quantum computing has recently garnered significant attention due to progress in building small-scale devices. However, significant technical advances will be required before a large-scale, practical quantum computer can be achieved.[26]

At present, then, the threat remains hypothetical. Those most familiar with the current research estimate that, assuming quantum computing is a real possibility, it will take at least a decade or two to develop it. But with the US and Chinese military establishments devoting massive scientific resources to its realization, it may well not remain hypothetical forever.

In Conclusion

As long as the NSA continues to operate without effective oversight, we must assume it is still listening in to our telephone conversations and reading our emails—albeit now with greater vigilance against whistleblowers like Edward Snowden.

Americans who feel and fear the constant gaze of authority upon them are not irrational. Cyberwar algorithmic surveillance technologies have been turned against us. Our private space has been invaded. Our whereabouts are being continuously monitored and our personal communications recorded. The Chinese government's Big Brother behavior is routinely described in the Western media as "authoritarian," "repressive," and "chilling."[27] Why do those words not apply equally to the United States?

CHAPTER 18

American Exceptionalism and the Ultimate Perversion of the Behavioral Sciences

AMERICAN COMMANDERS QUOTED in newspapers casually refer to their own forces as "the good guys" and whomever they are fighting as "the bad guys." Every American elected official at every level, every major party politician, every mainstream media editorialist and opinion columnist, is expected to concur. It is the central tenet of American Exceptionalism: America can do no wrong. The United States is the "leader of the free world," and all American wars are *just* wars in defense of freedom, democracy, and human rights.

The Ideological Curtain

American Exceptionalism is the ideological curtain behind which American science committed itself to enhancing the technology of war. A rigorous examination of the claim to national moral superiority is therefore essential to challenging the out-of-control militarization of American science.

For me and many other Americans who came of age in the 1960s, experiencing the blatantly unjust American war in Vietnam led first to questioning the essential goodness of all things American, and then to rejecting it. From the Vietnam era to the present, American Exceptionalism has increasingly been used to justify policing the world to preserve US corporate dominance of the global economic order.

The war in Vietnam ended when the Pentagon realized it could no longer depend on its fighting forces to fight. American GIs may not have fully understood why they had been sent to Southeast Asian jungles, but many knew it was a bad cause not worth dying for. African American soldiers in particular had been emboldened by Martin Luther King's condemnation of the United States government as "the greatest purveyor of violence in the world today,"[1] and Muhammad Ali's defiant refusal to fight the Vietnamese. In official parlance, this was described as a "morale problem." The antiwar GIs, however, were not demoralized—they had courageously rejected being cannon fodder in an imperialistic war of conquest.

America as a "Shining City on a Hill"

There are innumerable expressions of American Exceptionalism spanning the centuries to choose from, but a relatively up-to-date paean to the country's "exceptional role" by *New York Times* pundit David Brooks can serve as representative. "America," Brooks writes, was

> assigned by providence to spread democracy and prosperity; to welcome the stranger; to be brother and sister to the whole human race; and to look after one another because we are all important in this common project.[2]

Like many self-congratulatory national-origin myths, this one contains a kernel of truth. America was perceived early on by oppressed Europeans as a Shining City on a Hill, a refuge from persecution. And then the American Revolution did, as Abraham Lincoln famously declared, bring forth upon the North American continent a new and highly progressive form of political organization. It was not the world's first republic, but it was by far the largest, demonstrating that representative government was workable in major countries as well as in small city-states.

But Lincoln's claim that the new nation was "conceived in liberty and dedicated to the proposition that all men are created equal" was at best a half truth. The American victory over the British Empire was a decisive step forward in the worldwide struggle for democratic rights and freedom from despotism. At the same time, however, the great North American social experiment was deeply marred by two scarring circumstances of its birth that American Exceptionalism leaves out of account:

- the genocidal removal of the indigenous population to make way for the European settlers, and

- the use of millions of enslaved Africans as labor to create the nation's original accumulation of capital.

It was evident to all but the willfully blind that democracy, prosperity, liberty, and human rights were not meant for all people in the new republic. America as a Shining City on a Hill was never, right from the start, anywhere near as lovely as it appeared to be from a distance. The Trump administration's inhumane policies against refugees and immigrants have sullied its mystique to such a degree that political commentator Maureen Dowd lamented, "The shining city on a hill is an ugly pile of rubble."[3]

The Point of No Return for American Exceptionalism

In April 2004 an American prison in an obscure town in Iraq named Abu Ghraib became the focus of a worldwide scandal when revelations surfaced that US Army and CIA personnel were routinely torturing prisoners there. It was later revealed that medical and behavioral scientists were complicit in the cruelty.

Torture is the most unforgivable of all crimes against humanity. Since the eighteenth-century Enlightenment, all civilized societies have unequivocally condemned it as barbaric. It is a means no end can justify, and it is absolutely prohibited by international law. Today, any country guilty of practicing torture forfeits all claim to moral leadership.

Despite the Bush-Cheney administration's assurances that the events at Abu Ghraib were "isolated incidents" perpetrated by "a few bad apples," it soon became public knowledge that these crimes against humanity were commonplace and by no means confined to a single prison. Detainees were also being tortured elsewhere in Iraq and Afghanistan, and at the US Marine base at Guantánamo Bay, Cuba. An ACLU report describes the Guantánamo facility as

a laboratory for the torture methods associated with [the Bush-Cheney] administration's so-called war on terror and for improvised judicial proceedings that violate basic principles of due process and justice.[4]

Furthermore, American military and intelligence agencies had been caught conducting what they called "extraordinary rendition," wherein terrorism suspects were kidnapped and transported to black site prisons in secret locations overseas. There, where the rule of law was not even a pretense, they could be tortured with impunity and international human rights agencies could not bear witness.

It should never be forgotten that most of those who were "extraordinarily rendered" were merely *suspected* of involvement in rebellious activities. Many cases have been reported in the international press of young men randomly swept up in military raids or "sold to U.S. forces for bounty by Pakistani and Afghan officials, militia, and warlords."[5] Well-meaning human rights advocates have protested that these hapless torture victims were innocent, but guilt and innocence are not relevant considerations for people arrested in their own countries by an occupying army with no moral or legal right to be there in the first place.

The Torture Report

Growing international and domestic outrage finally forced American officials to directly address the torture accusations. An investigation was launched in 2009 under the auspices of the US Senate Select Committee on Intelligence (SSCI), which in December 2012 produced a report that made further denials impossible. It took two more years—until December 2014—before it was partially published, but by then public knowledge of its revelations had already prompted Barack Obama to concede that "we tortured some folks."[6]

The SSCI torture report was based on millions of pages of secret CIA documents, but almost ten thousand classified documents were withheld from the investigators by Obama administration claims of executive privilege. Despite that act of prior censorship, the report was apparently still so disturbing that less than 10 percent of it was made public.[7] Of the report's 6,770 pages, only a 525-page, heavily "redacted" summary was published.[8] The rest of the torture report remains classified, hidden from public view. The redacted summary's damning conclusions make it difficult to imagine what further revelations are concealed. Among the torture techniques recounted ad nauseum are beatings, sleep deprivation, prolonged subjection to extreme temperatures and "stress positions," confinement with insects, repeated forcible IV anal infusions called "rectal rehydrations," and repeated waterboarding.

Misbehavioral Science:
The Systematic Deconstruction of Human Minds

The last shreds of American Exceptionalism's legitimacy died at Guantánamo Bay, and scientists played a prominent part in shredding it. The watchdog organization Physicians for Human Rights charged that American physiologists, psychologists, and other medical specialists had approached the historic nadir of scientific morality attained by Nazi researchers during World War II:

> CIA contract psychologists James Mitchell and Bruce Jessen created a research program in which health professionals designed and applied torture techniques and collected data on torture's effects. This constitutes one of the gravest breaches of medical ethics by U.S. health personnel since the Nuremberg Code was developed in the wake of Nazi medical atrocities committed during World War Two.[9]

In 2007 the International Red Cross condemned the participation of medical personnel in CIA interrogations as "contrary to standards of medical ethics." CIA director Michael Hayden denied the charge in testimony before Congress, declaring that the CIA's physicians are present only "to ensure the safety and well-being of the detainee." The SSCI report refuted Hayden's assertion: "This testimony was incongruent with CIA records."[10]

That CIA doctors were not passive observers of the torture was further corroborated by agency documents declassified in 2016 following an ACLU legal challenge.[11] A medical ethicist reviewing those documents concluded:

> It's long been known that doctors attended torture as monitors. What's new is their role as its engineers. . . . CIA physicians played a central role in designing the agency's post-Sept. 11 torture program.[12]

Among other things, "doctors helped to design a waterboarding method more brutal than what even lawyers for the George W. Bush administration allowed." In doing so, those physicians "acted as a rogue profession, contemptuous of medicine's Hippocratic commitment to patients."[13]

But even if Hayden's claim had been true, it was irrelevant. Medical professionals monitoring the health of prisoners to assure that interrogators don't accidently kill them are as complicit in torture as the interrogators themselves. Their very presence violates the traditional Hippocratic injunction to "first, do no harm."

No-Touch Torture

No historian has investigated the CIA's research on interrogation methods more thoroughly than Alfred McCoy. "From 1950 to 1962, the CIA led secret research into coercion and consciousness," McCoy writes, which "produced a new method of torture that was psychological, not physical—best described as 'no touch' torture." The CIA's "new psychological paradigm," he adds, was "the first real revolution in this cruel science since the 17th century."[14]

> Although seemingly less brutal, no-touch torture leaves deep psychological scars. The victims often need long treatment to recover from trauma far more crippling than physical pain.[15]

The new paradigm is a two-stage method. The first phase is designed to disorient the prisoner by prolonged sensory and sleep deprivation. The images of hooded prisoners at Guantánamo and Abu Ghraib are the most recognizable representation of sensory deprivation. Once the prisoners have thus been "softened up," the second phase begins. That entails subjecting them to "self-inflicted discomfort"—for example, by making them stand with their arms extended for hours. The point is "to make victims feel responsible for their own pain and thus induce them to alleviate it by capitulating to the interrogator's power."[16]

The CIA distilled the fruits of its research into a handy how-to manual in 1963, which they shared with American military allies around the world.[17] The manual and its successors were at first distributed by the USAID Office of Public Safety and later, in the 1980s, by US Army Mobile Training Teams operating in Central America. USAID's involvement suggests that American "foreign aid" included the provision of torture-science textbooks.

Two Pioneers of Cold War Behavioral Science

"Though research into psychological torture was indeed a dark science," Alfred McCoy writes, "it was a science nonetheless and has a significant history that commands closer attention." He attributes its origins to "competing approaches by two titans of twentieth-century medical science—Donald O. Hebb's behavioral methods and Henry K. Beecher's drug experimentation."

> Their eminence should not be underestimated: Hebb is a towering figure in cognitive science, and his 1949 book *The Organization of Behavior* is regarded by some as second only to Darwin's *Origin of Species* in scientific significance; Beecher is remembered as a pioneer in American anesthesiology and, above all, modern clinical ethics.[18]

Donald Hebb bears primary responsibility for devising the CIA's sensory deprivation torture techniques, which were distilled from dangerous, damaging experiments on unwitting student subjects. His reputation as a hero of science was forged in an era when his unethical research, classified as a military secret, was hidden from public view. Today, despite the exposure of the dark side of his legacy, he is posthumously lauded as the "father of neuropsychology." A Donald O. Hebb Award in Behavioral and Cognitive Sciences is bestowed annually by the Canadian Psychological Association in his honor.

Henry Beecher is today renowned as a champion of medical ethics, for his fervent campaign against nonconsensual human experimentation. The story of how he got there, however, is bizarre. Before he became renowned as an ethicist, Beecher was among the least ethical of experimenters; in postwar Germany, he participated in CIA mind-control research that continued Nazi-era medical experimentation on human subjects.[19] But unlike classic sinner-to-saint narratives, this one had an odd twist. Rather than publicly repenting for his deeply blemished past practices, he tried to pretend they didn't exist. He believed, it would seem, that the research he'd done for the CIA would remain secret forever, but it did not.

Although his earlier misdeeds are now a matter of public record, his reputation as an ethicist remains intact. Like Hebb's, his name also adorns a prestigious award. The Harvard Medical School has presented an annual Henry K. Beecher Prize in Medical Ethics for more than two decades.

Bitter, Brutal Biscuits

The medical scientists most culpable in bringing the CIA's new "no touch" paradigm into the twenty-first century were the psychologists affiliated with a program called BSCT. The cute acronym, pronounced "Biscuit," stands for behavioral science consultation teams. BSCT was born in early 2002 but first came to public attention in 2004 during the outrage over Abu Ghraib. An International Red Cross report identified it as a group of psychologists, psychiatrists, and doctors working at the Guantánamo prison in close collaboration with military interrogators to construct "an intentional system of cruel, unusual and degrading treatment and a form of torture."[20]

The most notorious BSCT alumni are the aforementioned clinical psychologists James Mitchell and Bruce Jessen. Mitchell and Jessen were the most spirited evangelists of the gospel of scientific psychological torture in the post-9/11 era. They joined BSCT in 2002, but their part in developing the CIA's new torture paradigm was kept secret for several years until investigative reporter Jane Mayer identified them by name and described their activities in detail.[21] These revelations were corroborated by the SSCI torture report in 2014, and by further documents produced in discovery from an ACLU lawsuit against the two psychologists.

Mitchell's prior experience with torture came from his work in a military training program called SERE (Survival, Evasion, Resistance, Escape). SERE was designed to prepare US soldiers to resist torture if captured, but Mitchell "reverse engineered" it from a way to withstand torture into a way to inflict it. When FBI agents observing his techniques criticized them as ineffective, Mitchell responded, "Science is science."[22]

Mitchell and Jessen designed a suite of specific interrogation tactics for the CIA and carried out a research program to continually refine them. They also taught their torture regimen to interrogators, from Iraq to Guantánamo and all the black sites beyond. Nor were their contributions limited to the theoretical and pedagogical; they participated directly in "hands on" interrogations. They starred in some of the waterboarding videotapes made at a black site in Thailand under the supervision of Gina Haspel, whom Trump later appointed to head the CIA.[23]

The aura of scientific legitimacy the two PhD psychologists brought to activities widely perceived as war crimes was valued highly by the CIA. The dynamic duo was paid as much as $1,800 a day for their expertise, but that was just the beginning. In 2005 they formed a company, Mitchell, Jessen & Associates, which took in $81 million from the CIA over the following few years. After their public outing as America's foremost torture scientists, Dr. Jessen tended to lie low, but the undaunted Dr. Mitchell exploited his notoriety by marketing himself as a public speaker. He set the price for sharing his ruminations on "the minds of those trying to destroy America" at $15,000 to $25,000.[24] Another financial expression of the CIA's appreciation was $5 million worth of legal indemnification. That came in handy in 2015 when Mitchell and Jessen were sued by three of the victims they had tortured. The taxpayers paid for Mitchell and Jessen's lawyers' fees and court costs.[25]

APA Complicity in "Doing Special Things to Special People in Special Places"

The participation of professional psychologists in CIA interrogations had become a hotly debated issue by the time the American Psychological Association held its 2006 convention. A featured speaker, US Army surgeon general Kevin Kiley, made his position clear: "Psychology," he proclaimed, "is an important weapons system."[26]

Kiley's appeal aimed at influencing the APA's 150,000 members.[27] The association's primary leaders, already deeply involved in behind-the-scenes collusion with the Pentagon and the CIA, did not need to be convinced. By the 2008 convention, however, the APA membership had clearly rejected the national security justification for torture. A resolution was passed explicitly condemning "torture . . . under the direction and supervision of behavioral science consultation teams (BSCTs) that included psychologists."[28]

The evidence of the APA leaders' complicity was presented in a 2015 report by independent critics entitled *All the President's Psychologists*. A trove of internal emails from as early as 2003 between CIA psychologists and high-level APA officials proved both parties were aware that Mitchell and Jessen were "doing special things to special people in special places."[29]

Meanwhile, the APA's board of directors had authorized an even more extensive investigation into the matter, resulting in a 543-page report corroborating the critics' charges.[30] The APA officially acknowledged the "deeply disturbing findings" that included "troubling instances of collusion," issued an immediate apology, and promised effective measures to prevent future complicity with torturers.[31] The organization's sharp reversal reflected the outrage the great majority of American psychologists felt at their profession's covert foray into the moral abyss of the War on Terror.

I give the last word on this subject to the eminent American psychiatrist Robert Jay Lifton. The fundamental moral judgment on the direct line running from Dachau to Guantánamo has never been better expressed:

> Physicians and psychologists are members of healing professions, in which knowledge is expected to be applied in ways that combat illness and enhance human life. It is a special form of betrayal when such professionals dedicate themselves instead to destroying other human beings.[32]

*How
We Got
into
This
Mess . . .*

CHAPTER 19

The Explosive Birth of Big Science

ONCE UPON A TIME, science was mostly small-scale ventures carried out by inquisitive individuals with modest resources at their disposal. That age of innocent simplicity is long gone. Science today is the domain of large teams of professional researchers working on a grand scale with substantial governmental and corporate funding. How and when did Little Science become Big Science?

The transformation began around the turn of the twentieth century, as commercial enterprises in Germany created laboratories and hired talent in order to harness science to their industrial purposes. Government funding encouraged the growth of an academic-industrial research alliance. The example of German corporations like Bayer and Siemens inspired emulation elsewhere, including in the United States.

World War I stimulated the process, as German science introduced chemical weapons into warfare and other countries followed suit. After the war, organized science experienced another growth spurt as several German firms combined to create the world's largest chemical company, IG Farben.[1] Industrial laboratories also proliferated throughout Europe, North America, and Japan. By the beginning of World War II, they numbered in the thousands.

It was during that war that Big Science fully emerged onto the scene. At the outset, Germany was the world's number one scientific powerhouse; by the end of the war the German state, and the state of German science, lay in ruins.

226

The Rise of Big Science

Meanwhile, the exigencies of modern military conflict had compelled the American government to take command of the scientific enterprise. The centralized mobilization of research and the great increase in resources devoted to it produced a diversity of important results, from radar to antimalarial drugs, but its defining triumph was the nuclear fission bomb created by the Manhattan Project.

The creation of the nuclear weapon that destroyed Hiroshima and Nagasaki in August 1945 demonstrated the raw power of large-scale government-sponsored scientific research. For the next four and a half decades, the Cold War provided the rationale for a rapidly growing expansion of the US government's role in science. Glorification of Big Science by DuPont, Union Carbide, Lockheed Aircraft, and other mammoth corporations dominated the public discourse, and the stifling conformity of the McCarthy era allowed few dissenting voices to be heard. As he left office in early 1961, President Eisenhower famously warned of the dangers of a growing "military-industrial complex," but to little effect. His successor, John F. Kennedy, made no attempt to rein in the Pentagon as it prepared for all-out aggression in Southeast Asia.

The Manhattan Project

The Manhattan Project was established by the US Army to develop and build an atomic bomb in order to gain a decisive edge over the Axis powers in World War II. The US Department of Energy (direct descendant of the Manhattan Project) provides a thumbnail history of its progenitor:

> The Manhattan Project is the story of some of the most renowned scientists of the century combining with industry, the military, and tens of thousands of ordinary Americans working at sites across the country to translate original scientific discoveries into an entirely new kind of weapon. When the existence of this nationwide, secret project was revealed to the American people following the atomic bombings of Hiroshima and Nagasaki, most were astounded to learn that such a far-flung, government-run, top-secret operation existed, with physical properties, payroll, and a labor force comparable to the automotive industry. At its peak, the project employed 130,000 workers and, by the end of the war, had spent $2.2 billion.[2]

The Manhattan Project, it adds,

> became the organizational model behind the remarkable achieve-
> ments of American "big science" during the second half of the
> twentieth century. Without the Manhattan Project, the Depart-
> ment of Energy, with its national laboratories—the jewels in the
> crown of the nation's science establishment, would not exist as it
> does in its present form.

The subsequent growth of American science can be partially measured by the magnitude of the federal dollars devoted to it. After the war,
governmental R&D spending decreased, as was to be expected, but remained far above prewar levels. In 1947 it totaled $8.1 billion.[3] By 1990,
with the Cold War as the driving force, it had climbed to $112 billion.[4]
After a small, brief decline following the disappearance of the Soviet
Union and the end of the Cold War, the rise resumed, reaching $145
billion in 2016.

Big Science's Secular Patron Saint

The creation of Big Science in America is frequently attributed to Vannevar Bush, head of the Office of Scientific Research and Development,
which Franklin D. Roosevelt created in June 1941 to mobilize six thousand American scientists for wartime military research.[5] Vannevar Bush
(no relation to the later presidents Bush) was a scientific visionary. He
has been described as "one of the great overachievers of the 20th century,"
who "combined the skills of an engineer, a mathematician, and a scientist
with the organizational abilities of a successful military leader or company
president."[6] During the war, no individual wielded more influence over
American science. Most significantly, he persuaded Roosevelt to approve
the expedited creation of a nuclear weapon, thus bringing the Manhattan
Project into being.[7]

Near the end of the war, in 1945, Bush was part of a committee formed
to advise President Truman on use of the nuclear weapons at his disposal.
Many of the scientists who had created the atomic bomb urged Truman to
issue a warning to Japan before resorting to the nuclear option, but Bush's
committee disagreed. Its recommendation, which was accepted, was to
drop the bomb without warning at the first possible opportunity.

After the war, Bush turned his attention to how nuclear weapons could be eliminated in a world at peace. He was, in other words, out of step with the bellicose spirit of the Cold War, and his influence waned accordingly. Together with leading nuclear physicists, he advised Truman against the creation of the hydrogen bomb, but their wise counsel was not heeded.[8]

Although Vannevar Bush had been the most prominent advocate of government-sponsored Big Science, his vision did not materialize in the way he had hoped. Bush wanted federal funds to pay for large-scale research projects, but did not want political appointees, bureaucrats, or other nonscientists to have decision-making power over them. He wanted the advance of scientific knowledge to be directed by scientists.

The wartime experience, however, pointed in the opposite direction. The resources of science had been mobilized for military purposes and put in the hands of military officials, who had no intention of relinquishing their authority over them. Although federal science policy was nominally under civilian control, politicians would henceforth routinely defer to military interests. Big Science in the United States continued to be dominated in large measure by the war planners, and the scientists would remain subordinate to them.

The "Imperialism of Physics"

World War I has often been called the chemists' war. World War II was indisputably the physicists' war. German physicists gave their country the V-2 rocket that battered Britain, and American physicists ushered in the nuclear era. Their dominant role shaped the postwar development of American science in regrettable ways. One consequence has been an "imperialism of physics," in which physicists rule over all other disciplines.[9] Postwar governmental policy placed theoretical physics on a pedestal as the model science against which other fields are measured. As a result, a few "aristocrats of physics" emerged as the primary spokesmen for American science. An authority on science policy, Daniel Greenberg, has described how they arrogantly "implanted their values, including disdain for the social and behavioral sciences, on government science policy for decades."[10]

This problem stemmed not from the bigness of Big Science in itself but from Big Science's dependence on funding from the Pentagon and the intelligence agencies. The military men chose veterans of the

Manhattan Project—many with hardline Cold War views matching their own—as preferred collaborators. The physics profession thus came under the command of a conservative bureaucratic elite closely allied with the war machine.

Little Wheels Turning Bigger Wheels

Although scientists were not able to gain command of American Big Science as Vannevar Bush had wanted, they were able to influence its development because they were indispensable to its mission. The relationship between the scientists and the socioeconomic forces driving Big Science was mediated by a number of advisory committees and organizations. Think of the latter as the small gears that turn larger gears that power a machine.

One of the most important of the small gears at the base of Big Science's power train in the Cold War era was a clandestine group of scientists, mainly theoretical physicists, that operated under the name JASON. Among JASON's founders and advisors were the physicists John Archibald Wheeler, Charles H. Townes, Murray Gell-Mann, Hans Bethe, Edward Teller, William Nierenberg, Steven Weinberg, and Freeman Dyson. JASON helped shape governmental science policy in total secrecy; it was unknown to the public until 1971, when the historic leak of the Pentagon Papers revealed its existence.

Continuing the mechanical metaphor, the JASON wheel turned larger wheels by serving in an advisory capacity to two crucial institutions of Big Science: the RAND Corporation and DARPA, an agency within the Department of Defense. From its origins in 1960, JASON advised the military on physics-related problems, including cutting-edge nuclear-weapon and ICBM-detection technology. During the Vietnam War, JASON contributed to the development of an "electronic battlefield" by devising anti-infiltration technology to transmit signals from hidden sensors on the ground to aircraft and artillery. After the exposure of its support to the American offensive in Vietnam, JASON became a high-profile target of antiwar protest.[11]

In recent years, the influence of JASON has been displaced by the DSB, the Defense Science Board, an advisory group with a majority of members representing military contractors. As a former DARPA official declared in 2014, "The Jason scientists are hardly relevant anymore."[12] In 2019 the

Trump administration drove what may prove to be the final nail in JASON's coffin. A Pentagon letter to the group terminating its contract effectively "killed off all of Jason's work for defense and nondefense agencies alike."[13]

DARPA—Epicenter of the Military-Industrial Complex

Have you ever searched the internet for information? Used a GPS while driving, to find your way around? Asked Siri or Alexa a question? If so, your lifestyle has felt the influence of DARPA, the Defense Advanced Research Projects Agency. Those wonderful, amazing technologies—and many, many more—all began there. But DARPA is also responsible for much that has led to the tragedy of American science: killer drones, robot assassins, transhuman supersoldiers. Those inhuman technologies—and many, many more—also began there.

DARPA has two faces, one that smiles outward at the world from its website and YouTube channel, and another hidden behind a multilayered veil of official secrecy. DARPA's public relations face hails it as science's greatest gift to humankind, while the other conceals its principal role in creating futuristic technologies of death.

The agency's publicists never tire of reminding us that the prototype of the internet was called ARPAnet because it was created by DARPA's predecessor, ARPA. ARPAnet was a purely military project, and the internet was the most important of all "dual use" technologies—the many innovations that civilians as well as warriors can put to use. ARPA began as a somewhat panicky reaction to the Soviet Union's launch of Sputnik in October 1957. Its birth in February 1958 was not greeted warmly by the Pentagon chiefs and the spymasters, who at first viewed it as challenging their own control over scientific research. Eisenhower imposed ARPA on them as a way to cut through the obstacles presented by interservice rivalries. The name was changed to DARPA in 1972.

DARPA doesn't do research. Its job is to imagine new weapons and weapons systems and hire others to do the research that can transform DARPA brainstorms into reality. Then, if their innovative technologies pan out, the agency turns the projects over to the appropriate branches of the armed services or intelligence agencies. RAND and DARPA histories have often intersected and intertwined as RAND scientists performed research under DARPA contracts.

A Crucial Political Struggle that Shaped DARPA

Among those who had resisted the fledgling agency's creation were two of the most vociferous of the scientific establishment's warhawks, Ernest O. Lawrence and Edward Teller. Teller in particular took Cold War fanaticism to a sociopathic level.[14]

Teller is widely recognized in popular culture as the "father" of American science's most tragic consequence, the hydrogen bomb. He and Lawrence jointly played a prominent political role in the policy debate that ended in the decision to create it. After the Soviet Union tested its first atomic bomb in August 1949, pressure mounted within American policy circles to respond by producing a weapon a thousand times more powerful than the bombs that destroyed Hiroshima and Nagasaki.

Most of the leading nuclear physicists were horrified by the idea. Those who served on the General Advisory Committee (GAC) of the US Atomic Energy Commission, including its chairman, J. Robert Oppenheimer, unanimously declared that the H-bomb's indiscriminate destructiveness rendered it militarily worthless, capable only of serving a "policy of exterminating civilian populations."[15]

Teller and Lawrence, however, had previously waged an unrelenting battle against their colleagues' humanitarian qualms, and they lobbied hard in favor of the H-bomb. Their arguments owed less to science than to the anticommunist fever afflicting the country. Whether or not they actually influenced the final outcome, they were on the winning side. In January 1950, Truman gave the go-ahead to an expedited program to create the hydrogen bomb.

Oppenheimer lost more than the policy debate. Red-baited by Teller and others, he was permanently barred from government posts, a crippling blow to his scientific career. Other nuclear physicists, including those who had earlier shared Oppenheimer's opposition to the H-bomb, got the message and fell in line behind the Cold War militarists. As Oppenheimer's career collapsed, Teller's and Lawrence's ascended. In 1952, Lawrence became head of a new national nuclear weapons laboratory in Livermore, California, with Teller as his chief scientific advisor. After Lawrence's death in 1958, the facility would be renamed the Lawrence Livermore National Laboratory.

Although DARPA was born in the same year Lawrence died, he had a hand in shaping its character. His efforts to block its creation proved unsuccessful, but he maneuvered to have one of Teller's protégés, Herb York, appointed to a Pentagon position of command over the agency. After a couple of years, York was succeeded by another close Lawrence-Teller ally, Harold Brown, thus reinforcing the hypermilitarized trajectory of American science.

The Prognosis: Big Science Is Here to Stay

Major scientific contributions by future Gregor Mendels or Henrietta Swan Leavitts[16] may well continue to occur, but confirming their innovative insights will require Large Hadron Colliders, orbiting space telescopes, Human Genome Projects, and the like. Penetrating nature's most hidden secrets will increasingly depend upon large teams of scientists with very big budgets. No social change—not even transforming the global economic system from market-based to socialist planning— could return science to its premodern past.

Big Science could and should be a positive force for the pursuit of human happiness. What if DARPA, for example, were to direct its efforts and resources toward solving the crisis of global warming? Before answering that question, be careful what you wish for. In the current sociopolitical context, if DARPA were to tackle that problem, it would at best be seeking a technological fix consistent with American military interests. Early indications of its possible pursuits in that realm involved controversial interventions into the natural world that go by the name "geoengineering."[17]

"The last thing we need is to have DARPA developing climate-intervention technology," one proponent of geoengineering says. "Geoengineering is already so fraught with social, geopolitical, economic, and ethical issues; why would we want to add military dimensions?"[18] His point is well taken. Expecting DARPA to pursue science beneficial to humankind would be like expecting a chicken to lay a duck egg.

The more important point, however, is that if the direction of American science were removed from the militarists' hands, stopping and reversing global warming would become an achievable step toward a limitless horizon of human accomplishment.

CHAPTER 20

Operation Paperclip: The Nazification of American Science

THE SPACE AGE DAWNED on October 4, 1957, when Sputnik I was lifted into orbit by a Soviet R-7 Semyorka rocket. The USSR's creation of the first human-made Earth-orbiting satellite demonstrated the proficiency of Soviet science and triggered a frenetic Cold War rocket-science race.

As symbolically important as Sputnik was, however, American military strategists were far more distressed by something else the R-7 Semyorka had accomplished a few weeks earlier. On August 21, 1957, the Soviet missile had carried a payload more than thirty-seven hundred miles in a successful test flight, announcing the birth of the world's first ICBM. The Soviet Union had demonstrated the ability to deliver a thermonuclear warhead to Washington or New York.

But it was Sputnik that impacted the American public's imagination and provided the policymakers with the political backing they needed to pursue another Big Science effort on the scale of the Manhattan Project. Despite the Soviet Union's fast start in the rocket-science competition, the United States rapidly caught up with and surpassed its rival superpower. The USSR's first operational ICBM was deployed in 1958; the US followed close behind in 1959.

The Nazi Roots of American Rocket Science

The United States' ability to surpass the Soviet Union in missile technology owed a great deal to a secret program called Operation Paperclip that brought an influx of some sixteen hundred German scientists who had served Hitler's Third Reich. Many of these scientists had been active members—some in the upper echelons—of the Nazi party.

Some had created heinous weapons of terror for the Nazi military; others had performed lethal medical experiments on inmates at Dachau and other concentration camps. But rather than prosecuting these individuals for war crimes, Operation Paperclip sanitized their war records, gave them American citizenship and long careers in prestigious American research institutions, and rewarded them with accolades and high honors. The German émigrés were not a minor supplement to American science; they became its essential core.

Bringing Nazi war criminals to the United States was illegal under American law. But officials of the OSS (predecessor of the CIA), the FBI, and army intelligence—arrogating to themselves judgment of what the "national interest" required—secretly and knowingly broke the law. As investigative journalist Annie Jacobsen reports,

> The scientists who helped the Third Reich wage war continued their weapons-related work for the U.S. government, developing rockets, chemical and biological weapons, aviation and space medicine (for enhancing military pilot and astronaut performance), and many other armaments at a feverish and paranoid pace that came to define the Cold War.[1]

The presence of Nazi war criminals in American scientific programs became public knowledge early on. Popular disapproval was widespread. But despite vigorous protests by Albert Einstein, Eleanor Roosevelt, and other influential voices, Operation Paperclip was allowed to proceed under cover of official secrecy. The German newcomers were not, generally speaking, ostracized by their American colleagues. Even after the worst excesses of McCarthyism had been reined in, the Cold War justification that "if we don't use them, the Soviets will" was enough to intimidate most critics into silence. (The USSR's missile program also tapped the scientific talent of German war criminals, but the cream of the Nazi crop wound up in the United States.)

The details of Operation Paperclip only began to come into public view in the late 1980s. Another investigative reporter, CNN's Linda Hunt, was able to reveal some of those details by dogged pursuit of

Freedom of Information Act requests.[2] But the most important revelations came in 1998 with the passage of the Nazi War Crimes Disclosure Act, which declassified many of the Operation Paperclip files. Annie Jacobsen's 2014 book on the subject was the first thorough analysis of that "classified body of secrets and lies."[3] As Jacobsen observed, however, the full story of Operation Paperclip has yet to be told, because despite the 1998 declassification of eight million pages of documents, millions more remain classified to the present day.

The leaven of Nazi research and researchers provided by Operation Paperclip was undeniably a major factor driving the Cold War militarization of American science. Its legacy, Jacobsen demonstrates, includes ballistic missiles, sarin gas cluster bombs, and weaponized bubonic plague.

Wernher von Braun and American Rocket Science

If the secretive Operation Paperclip had wanted to publicize itself, its poster boy would have been German rocket scientist Wernher von Braun. During the war, von Braun headed the Nazi team that created the V-2 rocket.

V-2 stood for *Vergeltungswaffe Zwei* ("Vengeance Weapon Two"). An exhibit at the Smithsonian Air and Space Museum in Washington, DC, recounts the missile's impressive curriculum vitae: "The German V-2 rocket was the world's first large-scale liquid-propellant rocket vehicle, the first long-range ballistic missile, and the ancestor of today's large rockets and launch vehicles."[4] The museum's description provides noteworthy highlights of Vengeance Weapon Two's history. It was assembled by concentration camp prisoners, at least ten thousand of whom died in the process of manufacturing it—more people were killed building the rocket than being hit by it.[5] That statistic could not have brought much comfort to the families of the 7,250 British citizens killed by the fifteen hundred V-2s that battered London and surrounding areas in September 1944.

Operation Paperclip allowed von Braun to reinvent himself as a patriotic American rocket scientist and Cold War hero. The magnitude of the Nazi impact on the US missile program is reflected in von Braun's position as NASA's first director.

The popularity of a witty ditty about von Braun's Nazi past by political satirist Tom Lehrer demonstrated that not all Americans were taken in by the snow job. Two of its lines perfectly capture the essential amorality of Operation Paperclip:

> "Once the rockets are up, who cares where they come down?
> That's not my department!" says Wernher von Braun.[6]

Von Braun's spy-agency sponsors tried to whitewash his record to make him look like a pragmatic technocrat who had only grudgingly worn the Nazi uniform. That was a gross distortion of the truth, as Linda Hunt documented in 1985.[7] Von Braun had in fact been a high-ranking officer in the infamous SS, the Schutzstaffel, which ran the Gestapo, the concentration camps, and the genocidal extermination programs. He was deeply complicit in the deaths of the thousands of slave laborers who were killed producing his V-2 rockets at the Dora-Nordhausen concentration camp.

A former Dora inmate, Jean Michel, wrote a harrowing account of his own experience there. Its title succinctly connects the Nazis' rocket science to their war crimes: *Dora: The Nazi Concentration Camp Where Modern Space Technology Was Born and 30,000 Prisoners Died.*[8] The Dora laborers were terrorized into submission by seeing many of their fellow workers executed, dozens at a time, by hanging, directly above the V-2 assembly lines. Von Braun made ten official visits to the facility in winter of 1945. Did he give the orders for those executions? That is not known, but he certainly knew they were happening and didn't give orders to stop them.[9] He was, by any reasonable definition, a war criminal, but that did not prevent his appointment as head of NASA.

The critical importance of the Nazi physicists to American rocket science was made clear in a May 1945 dispatch from Major Robert B. Staver, a US Army intelligence officer who was acting as an Operation Paperclip talent scout:

> Have in custody over 400 top research development personnel of Peenemünde. Developed V-2. . . . *The thinking of the scientific directors of this group is 25 years ahead of U.S.* . . . Later version of this rocket should permit launching from Europe to the U.S.[10]

Staver was describing the ICBMs of the future. Indeed, less than a year later, when Wernher von Braun had become head of the American rocket program, he sent a memo to Robert Oppenheimer at the Los

Alamos nuclear weapons laboratory suggesting they combine their specialties in order to create a rocket to carry an atomic bomb. Later written up as "Use of Atomic Warheads in Projected Missiles,"[11] it envisaged a rocket capable of delivering a two-thousand-pound warhead to a target a thousand miles away.

Von Braun was far from alone among Operation Paperclip's recruits with deep ties to Nazi criminality. Some of the more noteworthy Nazis with major roles in creating the US missile program were Walter Dornberger, Arthur Rudolph, and Kurt Debus. Major General Walter Dornberger had been in charge of production of the V-2 rockets at Dora-Nordhausen. In the United States he became "America's mouthpiece for the urgent need to weaponize space."[12] Arthur Rudolph was a rocket scientist who had been the deputy production director at Dora-Nordhausen. His office in the factory had a window that looked directly out onto the assembly line above which executed laborers were hanged to terrify the workforce. At his death in 1996, Rudolph was commemorated in a *New York Times* obituary as "the developer of the giant Saturn 5 rocket, which launched a crew of American astronauts on the first manned flight to the moon in 1969."[13] Kurt Debus, like von Braun, was an SS officer, and he proudly wore his Nazi uniform to work. Early in his twenty-eight-year American career, Debus became the first director of the Kennedy Space Center. "There," his obituary tells us, "he directed the launch of the first American earth satellite, the first probe of the sun, and the first flight of primates. There he also oversaw the flight that made Alan B. Shepard Jr. the first American in space."[14]

Disguising ICBM Research as Space Exploration Research

Rocket science was not conceived for space exploration. It was for war. Furthermore, the weapons it produced were of little battlefield value; they were designed to wipe out civilian populations. None of this changed when the rocket scientists transferred their base of operations from Germany to the United States. American rocket science, however, has long been enveloped in the alluring aura of a noble quest to reach the stars. Many Americans—myself included—can't help but be fascinated by the idea of space exploration. Curiosity about "what's out there" excites our imaginations. Small wonder that our popular culture has long been awash in Star Trek and Star Wars fantasies.

Early in the Cold War era, American military and intelligence officials realized they could exploit the public's natural inquisitiveness for their own ends. The missiles they needed to carry warheads around the globe could be promoted as rockets to propel humans to the moon and beyond. Adding appeals to national pride made massive research budgets an easy sell. It was a bait-and-switch game: promise the moon rocket and deliver the ICBM.

The rocket men of Operation Paperclip were among the program's most vociferous cheerleaders. Von Braun "was ambitiously developing a persona for himself as America's prophet of space travel," says Annie Jacobsen. He and Dornberger "became national spokesmen on this issue."[15]

NASA, the "Space Rockets for Peace" Program

Investigation and utilization of the vast reaches of space beyond Earth's atmosphere were officially separated in 1958 with the creation of NASA. NASA would take charge of all "peaceful, scientific, and open" space research, and secret military space activities would remain with the Pentagon. The division of labor, however, was more cosmetic than real. Declassified documents obtained in 2014 by the National Security Archives at George Washington University revealed that providing "cover stories for covert operations" has been "part of the secret side of the National Aeronautics and Space Administration (NASA) since its inception." Although ostensibly "a purely civilian space agency," NASA was obliged "to commingle its activities with black programs operated by the U.S. military and Intelligence Community."[16]

US–USSR Space and Nuclear Treaties

The drive to militarize space went into sudden decline in 1967 when American and Soviet officials, fearing a runaway proliferation of space-based nukes, signed an Outer Space Treaty prohibiting the placement of weapons of mass destruction in Earth orbit, on the moon, or on any other celestial body. An additional 128 countries also joined the pact.

The treaty was rightfully hailed as a significant step back from the brink of global nuclear annihilation, but it also pointed to crucially important unfinished business. The Outer Space Treaty refuted the Cold War axiom that workable arms limitation agreements with an absolutely untrustworthy Soviet Union were impossible. What, then, continued to prevent the United States from entering into an international agreement to limit and eventually abolish all nuclear weapons?

Two modest steps in that direction were taken, first in 1972 with the Anti-Ballistic Missile (ABM) Treaty, and then in 1986 when Ronald Reagan and Mikhail Gorbachev signed the INF Treaty limiting the development of intermediate-range nuclear missiles. Both agreements, though limited in scope, were in effect for the duration of the Cold War and for many years afterward.

Unfortunately, however, in the twenty-first century the tide of history began flowing in the opposite direction. The Bush-Cheney administration, anxious to pursue "Son of Star Wars" research, withdrew the US from the ABM Treaty in 2002. More recently, on August 2, 2019, the Trump administration scrapped the INF pact altogether. The demise of the INF Treaty is expected to trigger a major escalation of the nuclear arms race, involving not only Russia and the United States but China as well.[17]

Trump's "Space Force" Initiative

Meanwhile, Trump also announced the creation of a new branch of the US armed forces that would remove responsibility for the militarization of space from the air force and transfer it to a newly minted space force.[18] The decision to create a US Space Force did not originate in Trump's lizard brain. It was an idea some aerospace corporations and warhawk politicians had been pushing for quite a while. Trump picked it up and echoed it.

James Mattis, who was Trump's secretary of defense at the time, declared, "We need to address space as a developing war fighting domain, and a combatant command is certainly one thing we can establish."[19] All of this serves to remove the last vestiges of pretense that vast expenditures on space science are for peaceful purposes of scientific exploration. Although the 1967 Outer Space Treaty is still nominally in effect, it has not prevented substantial military intrusions into space.

Astrophysicist Neil deGrasse Tyson has emerged in recent years as "Mr. Science" in the realm of popular culture. Tyson has written persuasively and passionately about the dependence of American astrophysics on the US quest for military dominance, without which "there would be no astronomy, no astrophysics, no astronauts, no exploration of the solar system, and barely any comprehension of the cosmos." Astrophysicists, he declares, "are the wagging tail on a large geostrategic dog."[20]

"Space," Tyson writes, "has been politicized and militarized from the opening moments of the race to reach it," but it didn't move "to the front lines of warfighting" until 1991, in the First Gulf War, which the US military dubbed "the first space war."

> Never before had a military force been so dependent on earth-orbiting satellites for extensive support of its war effort: strategy, tactics, planning, communication, identification of targets, weapons guidance, troop movements, navigation, long-range weather prediction.[21]

Physics Was Not the Only Science Defiled by Nazi Scientists

Treaties to limit the growth of arsenals of mass destruction are also relevant to another class of weapons—one that emerged from the medical sciences rather than rocket science. The former Nazi scientist who best personifies the link between the two fields is Dr. Hubertus Strughold, the "father of American space medicine."[22] Strughold, a physiologist, had organized and directed the Third Reich's medical research programs during the war. The Nazi high command was especially interested in discovering ways to give its troops advantages over enemy soldiers. To that end, extensive experimentation was conducted to quantify physiological limits of humans subjected to extreme battlefield conditions.

The locus of combat most dramatically affected by rapid scientific and technological advance was in the air. As jet engines replaced propeller-driven motors in warplanes, pilots and crews had to endure higher altitudes, lower-pressure environments, more extreme temperatures, more potent acoustic shocks, stronger *g*-forces caused by extreme acceleration, and other new physiological stresses. Strughold had been the director of the Luftwaffe's Institute for Aviation Medicine, which conducted extensive research using inmates from the Dachau concentration camp as subjects. The prisoners were subjected to inhumane experimentation that frequently ended in their deaths. Among other things, they were immersed in freezing water, forced to drink lethal amounts of salt water, starved of oxygen in low-pressure chambers, and exposed to malaria and mustard gas.[23]

Among the most notorious of the medical murder projects was one seeking ways to treat the extreme hypothermia experienced by Luftwaffe crews shot down over the frigid waters of the North Sea. Subjects of the "immersion-hypothermia" experiments conducted at Dachau in 1942 and 1943 were labor camp prisoners and Russian POWs. The procedures were described in a *New York Times* account:

> In the freezing experiments, victims were forced to remain naked outdoors or in tanks of ice water. Tests were made periodically as they were freezing to death and various efforts were then made to rewarm and revive them, [but] virtually all died.[24]

These experiments were conducted under the supervision of Dr. Siegfried Ruff, who reported directly to Strughold.

Another focus of Strughold's research was how hypoxia—oxygen deprivation—at high altitudes affects human beings. In a 1942 study, two hundred Dachau inmates were tested in a low-pressure chamber simulating altitudes of up to sixty-eight thousand feet. Eighty of the two hundred subjects died of asphyxiation, and the survivors were killed so their bodies could be autopsied.

Strughold's institute was also conducting research into desalinating saltwater. German airmen downed at sea with no potable water available could not resist drinking the seawater, causing them kidney damage and excruciatingly painful deaths. Experimental desalinization processes were tested on prisoners at Dachau, who were given saltwater intravenously and denied all food and drink. Few survived the ordeal.

There was an unbroken line of continuity between these secret Nazi projects and equally secret postwar work at the US Army Air Forces Aero Medical laboratories in Heidelberg, where Strughold served as co-director and recruited its researchers. For a year, from the fall of 1945 until September 1946,

> day in and day out, fifty-eight German physicians in white lab coats had been working on an array of research projects in state-of-the-art laboratories studying human endurance, night vision,

blood dynamics, exposure to bomb blast, acoustic physiology, and more. They all reported to Dr. Strughold.[25]

In 1947, Strughold and most of his team were transferred to the US Air Force School of Aviation Medicine at Randolph Field, Texas.[26] From that time until his death in 1986, he was lionized as a hero of American science. His wartime criminality, though long suspected, was carefully covered up for decades by Operation Paperclip and only fully revealed to the public posthumously.

Dr. Strughold was but one of many Nazi medical men who had spent the war years experimenting on helpless prisoners in concentration camps. Of all the reprehensible practices of Nazi science, these were the most heinous. Even more consequential, however, was the development of chemical and biological weapons for Hitler's war machine. Among the most culpable Nazi biochemical warfare researchers were Operation Paperclip alumni Dr. Kurt Blome, Dr. Walter Schreiber, and Dr. Otto Ambros. These men laid the scientific foundation of the American chemical and biological weapons programs.

Poison Gas, Nerve Agents, and Pathogens

Comparing degrees of inhumanity in the ways soldiers and civilians are killed in wars would in itself be an inhumane exercise. Nonetheless, the use of chemical and biological munitions excites a level of revulsion and abhorrence that attach to no other class of weapons. That is one of their attractions for the military mind: the widespread terror they provoke.

The modern history of such weapons began during World War I when, on April 22, 1915, the German army launched a massive chlorine gas attack against French troops at Ypres, killing five thousand and incapacitating ten thousand more. France and its allies responded in kind, and by the end of the war an estimated 124,000 tons of poison gases—mostly chlorine, phosgene, and mustard gas—had killed some ninety-one thousand soldiers.

On June 22, 1918, the United States entered the field with the establishment of the CWS, the army's Chemical Warfare Service. The CWS built major production plants at Edgewood Arsenal in Maryland, at Aberdeen Proving Ground, and at other sites. The Edgewood facility included three plants for the production of chlorine, phosgene, chloropicrin, and sulfur mustard gases. Between the World Wars, the CWS continued researching, developing, and stockpiling chemical weapons.

In 1942, President Roosevelt added biological warfare to the mix by approving the creation of a secret bioweapons program headquartered at Fort Detrick, Maryland. It was managed by the War Reserve Service, headed by George W. Merck, president of the Big Pharma firm that still bears his family name.[27] By 1945, both chemical and biological warfare had been placed under the aegis of the Chemical Warfare Service.

Meanwhile, German scientists in these fields had far surpassed their American counterparts. With regard to chemical weapons, the US military was still at the level the Germans had reached in World War I. By contrast, Nazi researchers had in the meantime discovered and developed a new, revolutionary, far more deadly class of chemical weapons: nerve agents. In 1936 an IG Farben chemist, Gerhard Schrader, was experimenting with molecules that combined phosphorus and cyanide when he discovered a particularly potent compound.[28] IG Farben suspected the German military might be interested. They were.

The new compound, the first of the nerve agents, was named tabun. Exposure to tabun killed in a matter of minutes, while the old-fashioned World War I gases took hours. Furthermore, tabun didn't have to be inhaled by its victims in order to kill them; a droplet on the skin would do the job. Gas masks offered no protection.

Nazi military researchers immediately set about weaponizing the poison. Tabun-filled projectiles were mass-produced as artillery shells and aerial bombs designed to explode and vaporize the tabun into a highly toxic mist. By 1943, tabun production levels at the slave-labor Dyhernfurth factory had reached 350 metric tons per month; by the end of the war 12,000 metric tons had been stockpiled.

In the meantime, Gerhard Schrader had not been resting on his laurels. By 1939, he and his colleagues had invented another nerve agent, sarin, which was even deadlier than tabun. When sarin proved to be superior to tabun in other ways, new facilities were built for its mass production as well. Although never officially recruited into Operation Paperclip, Schrader nonetheless collaborated with postwar US military efforts at the Aero Medical laboratories in Heidelberg to develop sarin production.

By the end of the war, the Chemical Warfare Service officials were planning to secretly bring the Nazi chemists to the United States. Among those highest on their wish list was Otto Ambros, a codiscoverer of sarin and a leader of the efforts to weaponize nerve agents for the Nazi war machine. During the war, Ambros managed IG Farben's slave-labor chemical factory at Auschwitz. Birkenau, the Nazis' largest extermination facility, also happened to be at Auschwitz. The two were not unrelated. IG Farben's poison gases claimed more than a million victims in Birkenau's gas chambers.

After the war, Ambros was tried at Nuremberg and convicted of war crimes. He received a remarkably light sentence, eight years in prison, and served only two and a half. In early 1951 the JIOA,[29] the US agency overseeing Operation Paperclip, put Ambros on a list for accelerated treatment even though he was still in prison at the time. Within days he not only was freed but also had his considerable wealth restored to him. Due to his status as a convicted war criminal, the JIOA was unable to bring Ambros into the United States, but that did not prevent his being hired to work at Camp King, a secret US Army research facility at Oberursel, in the American sector of occupied Germany. Ambros later carved out a lucrative career as a scientific advisor to major American chemical firms such as Dow Chemical and W. R. Grace.

Meanwhile, the Nazis' success in weaponizing disease pathogens had not escaped the attention of the Chemical Warfare Service. The biochemist they coveted above all others was Kurt Blome. Deputy surgeon general of the Third Reich and a lieutenant general in the *Sturmabteilung* (known as the SA), Blome had been in charge of the Nazis' biological weapons research. He was known to have been working on a bubonic plague weapon, utilizing live plague pathogens in his research, and was suspected of testing them on concentration camp prisoners.

At Nuremberg, irrefutable evidence showed that Blome had planned extensive experimentation on human subjects. His defense was that, sure, he had planned to commit those atrocities, but never actually did. He was acquitted because no definitive evidence was found that could place him at the scene of the experiments or show him issuing orders for them. Despite the acquittal, US law nonetheless prohibited Blome's employment in any capacity by the United States. But US bioweapons officials desperately wanted to acquire his unique expertise, so Blome, like Ambros, was quietly signed to a contract to work at Camp King.

Yet another high-ranking Paperclip recruit at Camp King was Dr. Walter Schreiber. While Blome's wartime responsibility had been to create bioweapons, Schreiber's had been to create defenses against them. That does not mean his research was more benign than Blome's. Defensive technologies are not passive elements of warfare; they are no less essential to military aggression than their offensive counterparts. Without the shield a warrior dares not wield the sword.

Schreiber, as surgeon general of the Third Reich, was above Blome in the Nazi medical hierarchy—he was the German army's chief medical scientist. Despite being strongly suspected of war crimes, he, too, was eagerly sought out by Operation Paperclip recruiters for his expertise in bacteriology and epidemiology. After his stint at Camp King, he joined Dr. Strughold in Texas at the laboratories of the School of Aviation Medicine.

Dr. Schreiber's wartime activities, however, were eventually exposed by a courageous and persistent American war-crimes investigator, Dr. Leo Alexander. Evidence demonstrated that Schreiber had overseen "yellow fever experiments, epidemic jaundice experiments, sulfanilamide experiments, euthanasia by phenol experiments and the notorious typhus vaccine program 'with its 90% death rate.'"[30]

Schreiber's career in American science came to a rapid halt in the early 1950s when these revelations found their way into the American press.[31] He avoided being put on trial for war crimes, however, by agreeing to leave the United States voluntarily. He found asylum—and was able to continue conducting medical research—in Argentina, where he lived until his death in 1970.

Schreiber's personal downfall did not seriously impede the process of technology transfer from Germany to the United States. The development of biochemical weapons begun by Nazi scientists continued under the guidance of many of the same scientists, who in the Cold War context professed themselves to be proud, patriotic Americans.

Toward the end of the 1960s, the Pentagon was spending hundreds of millions of dollars a year building a formidable stockpile of weaponized nerve agents and pathogens. Before the decade ended, the United States would declare that it was ceasing production of chemical weapons and destroying its stockpile of biological weapons. In the meantime, however, military researchers had performed a great deal of bioweapons experimentation on the American public without its consent or even knowledge.

Germ Warfare Tests: Operation Sea Spray and the New York City Subways

For six days at the end of September 1950, the US Navy secretly sprayed a cloud of mist containing *Serratia marcescens* bacteria into the air just off the coast of San Francisco as the wind was blowing ashore. They called it Operation Sea Spray, and its purpose was to assess the vulnerability of a large coastal urban area to a biological warfare attack.

Hundreds of thousands of residents of San Francisco, Berkeley, Oakland, and other Bay Area towns were unaware of being guinea pigs in an experiment that caused them to inhale millions of the bacterial spores. The researchers who designed the tests believed *Serratia marcescens* posed no threat to humans. They were wrong. Today it is recognized as a harmful pathogen that can "cause a wide range of infectious diseases, including urinary, respiratory, and biliary tract infections."[32]

Operation Sea Spray was part of a top-secret military bioweapons program that comprised at least 239 experiments in at least eight American cities from 1949 to 1969. Another troubling case involved a large-scale release of bacteria in the New York City subways in 1966. On June 6 of that year, twenty-one US Army researchers smashed light bulbs on ventilation gratings above the 6th Avenue and 8th Avenue subway lines in Manhattan. Each light bulb contained approximately 87 trillion *Bacillus globigii* bacteria. After the microbial spores were dumped into the 23rd Street station, air currents stirred up by the moving trains circulated them from 14th to 59th Streets. Over the next five days, the army reported, more than a million subway riders had been exposed to the "harmless" germ.[33]

Once again, it was not harmless. *Bacillus globigii*, a frequent cause of food poisoning, was later categorized as a pathogen. Although a National Academies investigation found that the "infections are rarely known to be fatal," people have been known to die from them.[34]

Operation Sea Spray and the New York City subway experiments were not revealed to the American public until 1976 and 1975, respectively.[35] Meanwhile, a dramatic reversal in biowarfare policy had occurred.

Outlawing "the Poor Man's Atom Bomb"

In 1969, Richard Nixon declared a unilateral moratorium on the United States' production of chemical weapons and possession of biological weapons. That such an announcement would come from a man who was at the time prosecuting a merciless war of aggression in Southeast Asia may seem paradoxical.

There was a logic to Nixon's opposition to biochemical weapons. As senator John Kerry would later explain, for countries determined to produce weapons of mass destruction, "chemical weapons are the most financially and technically attractive option."

> The ingredients for chemical weapons are chemicals that are inexpensive and readily available in the marketplace, and the formulae to make nerve and blister agents are well known. It is no coincidence that chemical weapons are known as the poor man's atom bomb.[36]

The year was 1997. Nixon's moratorium had expired a decade earlier, and Kerry was arguing that the United States should ratify the Chemical Warfare Convention, or CWC, which bans the development, production, stockpiling, and use of chemical munitions. The horrific use of chemical weapons in the 1980s by Iraq against Iran and against its own Kurdish minority had given new urgency to international efforts to outlaw them. The CWC was ratified by the United States, and, as of 2018, 192 other countries were also parties to the agreement.

As for germ warfare, the United States had earlier joined in a similar Biological Warfare Convention, or BWC, to which (also as of 2018) 181 other countries had signed on. That pacts of this sort cannot guarantee successful arms control was dramatically demonstrated in early 2018. The CWC's watchdog arm, the Organization for the Prohibition of Chemical Weapons, accused one of its member states, Syria, of using nerve agents against its own rebellious population. Syria denied the charges, and an ensuing OPCW vote to censure Syria was "largely symbolic."[37]

Despite alleged Syrian violations, the treaties have succeeded to some degree in restraining use of the heinous weapons. American policymakers back diplomatic efforts to ban "poor man's nukes" because biochemical warfare is an erratic, unpredictable game in which the United States has minimal advantage. Meanwhile, the US arsenal of "rich man's nukes" continues its relentless advance.

The Nazi Nexus in the Behavioral Sciences

Rocketry, physiology, and biochemistry were not the only American sciences contaminated by the moral legacy of Operation Paperclip. The behavioral sciences were no less compromised. A 1952 CIA memo reveals the agency's keen interest in their potential. Psychiatry and psychotherapy, it says, have

> developed various techniques by means of which some external control can be imposed on the mind/or will of an individual, such as drugs, hypnosis, electric shock and neurosurgery.[38]

There was no need for American science to start from scratch in this field, because a solid foundation had already been laid by German researchers during World War II:

> Just as the U.S. space program benefited from the work of Werner von Braun's rocket scientists at Peenemünde and the Luftwaffe's murderous medical experiments at Dachau, so this CIA mind-control effort continued the research of the Nazi doctors, both their specific findings and their innovative use of human subjects.[39]

A large cadre of American behavioral scientists abetted this continuity, the most important of whom was the anesthesiological pioneer Henry K. Beecher.[40] In 1947 the head of the army's Medical Intelligence Branch sent Beecher a brochure describing thirty experiments performed on Dachau inmates. The research was designed to assess the value of "ego-depressing" drugs, especially mescaline, in interrogating prisoners. Four years later, Beecher embarked on an extensive European tour to pursue that subject. It led him to the former Nazi interrogation center at Oberursel, which under American management had changed its name to Camp King. "There," historian Alfred McCoy writes, "a staff of ex-Gestapo soldiers and former Nazi doctors, including the notorious deputy Reich health leader Kurt Blome, were employed in inhumane interrogations of Soviet defectors and double agents."[41]

Beecher also collaborated at Camp King with another Operation Paperclip notable, Dr. Walter Schreiber, whom he described as "intelligent and helpful."[42]

The Abominable Dr. Gottlieb

Sidney Gottlieb was a chemist, not a behavioral scientist. His role in CIA mind-control experimentation, however, was no less odious than Beecher's.[43] Gottlieb was a central figure in the CIA's insidious MK-ULTRA program that sought to use mind-altering drugs such as LSD to manipulate the human psyche.

A new biography by journalist Stephen Kinzer describes Gottlieb's activities in the early 1950s in macabre detail.[44] In a secret CIA detention center in Munich, he drugged prisoners of war, interrogated them, and left them to die. Gottlieb and his team called their experimental subjects "expendables" because they could dispose of their bodies with no consequences.

In an interview, Kinzer explained that MK-ULTRA was essentially a continuation of Japanese and Nazi research. "The CIA," he said, "actually hired the vivisectionists and the torturers who had worked in Japan and in Nazi concentration camps to come and explain what they had found out so that we could build on their research."[45]

Considerations of morality aside, even on its own terms MK-ULTRA's brainwashing research was scientifically worthless. As Gottlieb himself ruefully admitted at the end of his career, he never succeeded in controlling any minds.

Shiro Ishii's Chamber of Horrors

The Japanese contribution to American science cited by Stephen Kinzer should not be overlooked. Although the inhumane practices of the Nazi researchers have gained far more notoriety, their Japanese counterparts were no less sadistic and cruel. When postwar American policymakers became aware of Japanese germ warfare research, they shielded the researchers from prosecution as war criminals because they coveted the science their labs had produced.

The primary Japanese biowar facility, a site in Manchuria called Unit 731, employed some three thousand scientists. They were commanded by Dr. Shiro Ishii, an army surgeon with the rank of general. The army supplied Dr. Ishii with thousands of captured Chinese soldiers and Japanese dissidents to use as experimental subjects.

Warning: proceed with caution. Kinzer's account of the experiments at Unit 731 can induce nausea and recurring nightmares. I reproduce it here as a concise summary of extreme scientific depravity that must never be erased from the historic record. The prisoners at Unit 731

> were exposed to poison gas so that their lungs could later be removed and studied; slowly roasted by electricity to determine voltages needed to produce death; hung upside down to study the progress of natural choking; locked into high-pressure chambers until their eyes popped out; spun in centrifuges; infected with anthrax, syphilis, plague, cholera, and other diseases; forcibly impregnated to provide infants for vivisection; bound to stakes to be incinerated by soldiers testing flamethrowers; and slowly frozen to observe the progress of hypothermia. Air was injected into victims' veins to provoke embolisms; animal blood was injected to see what effect it would have. Some were dissected alive, or had limbs amputated so attendants could monitor their slow deaths by bleeding and gangrene.[46]

The American officials of the Biological Warfare Laboratories at Fort Detrick wanted to learn everything they could from the Japanese scientists who performed these atrocities. When arrested in early 1946, Ishii offered his captors a deal: "If you will give me documentary immunity for myself, superiors, and subordinates, I can get all the information for you." Fortunately for Ishii, the Supreme Command in occupied Japan had covertly decided that "the value to the U.S. of Japanese biological weapons data is of such importance to national security as to far outweigh the value accruing from war-crimes prosecution."[47]

The Supreme Commander, General Douglas MacArthur—with approval that went all the way up the chain of command to Harry Truman—secretly granted amnesty not only to Ishii but to all of his colleagues at Unit 731. Unlike the Operation Paperclip scientists, however, they were not transported to the United States. Instead, laboratories were set up for them in East Asia so they could quietly continue their research under American supervision.

The officially orchestrated cover-up of these horrendous crimes succeeded in hiding them from public view for more than forty years. Not until the mid-1980s did detailed, documentary evidence of what had transpired at Shiro Ishii's Unit 731 begin to emerge. But subsequently, an American journalist noted, "a trickle of information about the program has turned into a stream and now a torrent."[48] The resulting histories have rendered the continued denials by Japanese officials simply preposterous.[49]

The Operation Paperclip Balance Sheet

In the final reckoning, how did Operation Paperclip impact the development of American science? Did infection with the Nazi virus render science in the United States as morally degenerate as that of its Third Reich forebear?

The theological doctrine of original sin is not helpful in understanding the descent of American science into its current tragic condition. The Nazi immigrant scientists did not control or determine the course of American science. That blame lies with the United States' own Cold Warriors. The amoral ethos the ex-Nazis brought with them, however, fed and encouraged the Cold War militarism in the United States that continues to define the direction and character of American science today.

CHAPTER 21

The RAND Corporation: From "Fuck You, Buddy" to Doomsday

FOLLOWING WORLD WAR II, A SYSTEMATIC APPROACH to formulating policies guiding military-industrial Big Science evolved in think tank form as the RAND Corporation. The acronym is a mash-up of "R and D." Some of the titles of books and articles about RAND provide a hint of what it represents in the public imagination: "The Think Tank that Controls America," "America's University of Imperialism," "Dr. Strangelove's Workplace," "Wizards of Armageddon."

RAND's early years were shrouded in official secrecy as the research it performed was almost entirely classified. As former RAND associate Chalmers Johnson explained, RAND "has always relied on classifying its research to protect itself, even when no military secrets were involved."[1] Nonetheless, this much we do know. It originated in October 1945 as the Rand Project within the Douglas Aircraft Corporation. The air force brought it into being in collaboration with the US Department of War (soon to be renamed "Department of Defense") and Vannevar Bush's OSRD. Its mission was to coordinate federal and corporate R&D with military planning in order to ensure American technological preeminence. In 1948 it formally separated from Douglas Aircraft and became the RAND Corporation, a private nonprofit organization.

A few of the seminal contributions to the development of high-tech warfare that RAND claims were:

- **The internet:** RAND engineer Paul Baran made important advances in the technology of packet-switching, which DARPA

253

helped the Pentagon develop into its worldwide ARPAnet system, the predecessor of the internet.

- **Supercameras on spy satellites:** RAND's Corona Program produced a satellite reconnaissance system for photographic surveillance that "became the backbone of American intelligence on the Soviet Union. They watched troops march along the Russian border with China and spied on cities they'd never seen before. They could even count the fruit in Soviet orchards and analyze their crops."[2]

- **ICBMs and MIRVs:** RAND physicist and mathematician Bruno Augenstein "was widely regarded as the father of the American intercontinental ballistic missile."[3] He also invented the Multiple Independently-targeted Reentry Vehicle, which gave ICBMs the capability of carrying several thermonuclear warheads that could each be aimed at a different target.

- **The neutron bomb:** RAND physicist Samuel Cohen invented this tactical thermonuclear weapon that "maximizes damage to people but minimizes damage to buildings and equipment."[4]

But the RAND Corporation's signature initiative, and perhaps the most consequential, was its hypertheoretical venture into "thinking the unthinkable." Its legendary "Nuclear Boys Club" waded into the topsy-turvy world of megatons and megadeaths in a quixotic effort to put nuclear warfare strategy on a rational, scientific basis.

Deterrence Theory: A "Game" of Global Risk and Peril

In 1945, when American policymakers made the decision to drop atomic bombs on Japan, the game was simple. They did not have to fear retaliation because they knew Japan could not respond in kind. The more farsighted among them were aware, however, that the United States' monopoly of nuclear weapons could not last forever.

The first explosions of the bombs, in tests in New Mexico and then lethally in Japan, exposed the most important nuclear secret for the world to see: *It works!* Nuclear fission chain reactions were not the hypothetical fever dreams of mad scientists, but reality. The knowledge that atomic bombs were possible made it inevitable that physicists in other parts of the world would sooner or later figure out how to make them.

American policymakers had no choice but to adjust to that reality. A debate among them ensued over how best to assure that the fearsome power they had unleashed upon the world would not be turned back against the United States. Their deliberations posed the complex problems that the RAND Corporation's nuclear strategists took on. Some of the policymakers clung to the unrealistic hope that strong enough security measures could be put in place to safeguard nuclear technology and keep it out of hostile hands indefinitely. Wiser heads knew better, and suggested instead an attempt at managing the proliferation of nuclear science by sharing it with allies to offset its acquisition by enemies. No one, however, really believed that would be an adequate solution to preventing a nuclear World War III.

In fact, following the dropping of the bombs on Japan, it was only four years before the Soviet Union had become the second member of the nuclear club. On August 29, 1949, the USSR exploded its first nuclear weapon at its Semipalatinsk test site in Kazakhstan. This was *(ahem)* a game-changer.

The Soviet nuclear test in August 1949 prompted the United States to immediately up the ante. President Truman announced the intention to create a thermonuclear weapon, or "superbomb," with explosive power that would dwarf that of the bombs dropped on Japan. In response, the Soviet Union vowed to do the same. An arms race to Armageddon was under way.

The United States performed its first hydrogen bomb test in November 1952, releasing the energy equivalent of 10.4 megatons of TNT—more than seven hundred times that of the bomb that destroyed Hiroshima. In less than a year, in August 1953, the USSR replied with a much smaller (400 kiloton) fusion device test, but it was enough to demonstrate that their scientists had learned how to make hydrogen bombs. In 1961 the Soviet Union tested the largest bomb ever detonated—a 50-megaton behemoth with the explosive force of 3,800 Hiroshima blasts.

The Cold War competition begat unrestrained growth of thermonuclear stockpiles. At their peaks, the US arsenal contained 31,255 nuclear weapons and the Soviet Union's held 40,159.[5] These numbers declined after the Cold War ended, but, today, they still represent enough firepower to obliterate the human population of the planet Earth many times over.

Furthermore, their proliferation to other nations makes the danger they pose now greater than ever before. As of 2013, nine nations had more than ten thousand nuclear warheads in their arsenals, not counting additional thousands of "retired" warheads in storage.[6]

Planning to Win the Game that Ends in Doomsday

Early in the course of the Cold War, American policymakers saw the world in stark terms: two implacable superpowers face-to-face, armed to the hilt with weapons on hair-trigger alert that could conceivably destroy all human life. What could be done—what steps should be taken—to prevent the world from ending in thermonuclear conflagration?

General Curtis LeMay, as head of the Strategic Air Command, was in charge of the US nuclear strike forces. His proposal typified the kill-it-in-the-cradle instincts of the military mind: He called for massive preemptive nuclear strikes against the USSR. He advocated that policy despite the Strategic Air Command's own estimates that it would anni-hilate more than seventy-seven million people in 188 targeted cities.[7] Fortunately, President Eisenhower had the final word and ruled against preemptive strikes.

The official nuclear policy of the Eisenhower administration, as articulated by secretary of state John Foster Dulles, was the doctrine of "massive retaliation," an explicit warning to the Soviet Union that the United States would not hesitate to meet any Soviet act of aggression by unleashing the full force of its nuclear arsenal.[8] The bigger the threat, it was assumed, the more effective the deterrent.

Meanwhile, the RAND Corporation's "national security analysts" had taken on the assignment of thinking more deeply about the problem and promptly identified fatal flaws in the massive retaliation strategy. These men—"eggheads in a world of meatheads," in the words of intellectual historian Louis Menand—"regarded the uniformed military man" as "a relic of the pre-scientific dark ages."[9] The RAND policy wonks were the polar opposite of their military counterparts. Their direct experience with war was negligible. They were mainly physicists, mathematicians, and engineers, with an occasional economist or political scientist added to the mix. When critics called attention to their lack of practical military experience, they replied that nobody had experience with the kind of all-out nuclear warfare that needed to be analyzed and understood.

RAND's war strategists acquired a number of colorful nicknames, including the Nuclear Boys Club and the Megadeath Intellectuals. The most famous member of the group was an ebullient grandstander named Herman Kahn, "a jocular, gregarious giant who chattered on about fall-out shelters, megaton bombs, and the incineration of millions."[10]

Herman Kahn: "On Thermonuclear War"

Kahn, a physicist who joined RAND in 1947, worked with Edward Teller, John von Neumann, and others on the project that produced the American hydrogen bomb. He gained celebrity status, however, as a tireless popularizer of doomsday prophecy. His provocative, whimsical ruminations about the end of the world made him a target of Stanley Kubrick's wickedly satirical *Dr. Strangelove*.[11] Kahn of RAND was not Kubrick's only inspiration for Strangelove of BLAND. Others included Teller, Wernher von Braun, and Henry Kissinger, but Kahn was certainly prominent among them. Far from being offended by the parody, Kahn reveled in it. He enthusiastically embraced the public role of provocateur scientist jolting the masses out of their foolish complacency.

Kahn presented his views in a 1960 book entitled *On Thermonuclear War*. He insisted, first of all, that an all-out nuclear war between the superpowers was preventable. But because that could not be guaranteed, the next question was whether the United States could *win* such a war. He concluded that it could.

But that, too, was not a sure thing, so the final consideration was: Could it at least be *survived?* Once again, his answer was yes, and he offered a strategy for this three-stage scenario.

To begin with, Kahn dismissed the massive retaliation strategy as unhelpful. It creates an unstable confrontation, like two wild-west gunslingers pointing their six-shooters at each other at point-blank range. Why would either of them hesitate to pull the trigger? And what could be less desirable than an unstable nuclear faceoff?

A key step toward stabilizing such confrontations, said Kahn, is to gain *second-strike capability*. If Gunslinger A can convince Gunslinger B that he (Gunslinger A) will be able to survive at least the first volley of bullets fired at him and will be able to return deadly fire, then Gunslinger B will have good reason not to fire first. If both gunslingers are assured of the other's second-strike capability, the result should be a stable standoff—*theoretically.*

In nuclear confrontations, credible second-strike capability requires having such a massive arsenal that even if a large portion of it is destroyed in a first strike, enough missiles with nuclear warheads will remain to kill millions of the first-strikers' citizens. Stability thus arises from a scenario wherein neither side has reason to believe they could destroy their enemy without putting themselves in grave danger. To achieve that, both sides must amass such fearsome nuclear arsenals that both are afraid to use them.

The quest for security in a "balance of terror," however, was illusory. What actually occurred was a frenzied nuclear arms race. While not exactly what Kahn had in mind, the official US deterrence strategy became known as MAD, or Mutually Assured Destruction. MAD, most independent commentators concluded, was indeed madness.

The Strangelove Paradox:
"The Whole Point Is Lost if You Keep It Secret!"

There is an ironic twist in the story of deterrence theory that reveals one of the deepest dimensions of the tragedy of American science. It has to do with something called "the Strangelove Paradox." The science of nuclear strategy pioneered by RAND produced pristinely rational and logical policy recommendations. Deterrence depended upon two iron principles. First was the above-mentioned second-strike capability: the United States must have the capacity to withstand any enemy's first nuclear strike and still be able to respond with overwhelming force. Second—and this was absolutely crucial—the enemy must be *made to know*, with *no ambiguity whatsoever*, that the United States will be able to carry out that deadly counterstrike. The logic is impeccable. It is that knowledge that forestays the enemy's hand.

Dr. Strangelove, in the movie, explained this point with regard to his Doomsday Machine's deterrent value: "The whole point of the Doomsday Machine is lost if you keep it a secret." If the other side does not know it exists, it can't deter anything. To be effective, it has to be publicized, not hidden.

As it turned out, RAND's elegantly logical deterrent strategy was too subtle for the secrecy-obsessed military mind to grasp. American military policymakers utterly missed the crucial point. They devised plans to assure their ability to retaliate . . . *and kept them secret!* Furthermore, their troglodyte counterparts in the Soviet Union followed the same illogical strategic course. That, then, is the final irony in the story of RAND's vaunted science of nuclear strategy. All of their sophisticated wrongheadedness with regard to nuclear security went for naught, because the policymakers on both sides steadfastly ignored their counsel. Our planet, it seems, has avoided nuclear holocaust until now less by shrewd deterrence stratagems than by dumb luck.

A Beautiful Mind?

One of several boldface names among the RAND Corporation's nuclear analysts was John Nash, a gifted mathematician who played a crucial role in creating the mathematical framework that underpinned RAND's deterrence strategies. His descent into paranoid schizophrenia was the subject of the Oscar-winning film *A Beautiful Mind*.[12] Paranoid schizophrenia is a devastating mental illness whose victims deserve the utmost compassion, but it is terrifying to contemplate that the fate of the human race could rest on a framework built upon paranoid delusions.

Nash was a pioneer of mathematical game theory, which analyzes the rules of games in order to devise winning strategies. Among its theoretical progeny was James M. Buchanan's public choice theory, which attempts to rationalize extreme economic inequality.[13] Nash's even more noxious application laid the foundation of Cold War nuclear brinkmanship policies.

Mathematical game theory was not the brainchild of Nash alone. Other pioneers were John von Neumann, Oskar Morgenstern, Thomas Schelling, and Albert Wohlstetter. Von Neumann and Morgenstern inaugurated the field with a 1944 book entitled *Theory of Games and Economic Behavior*. It was a bold attempt to put economic theory on a rigorous, axiomatic base derived from mathematical models of economic decision-making. Professional economists at first ignored game theory, but other social scientists began to notice its potential applications in other fields. Most significantly, it was picked up and transformed by the Nuclear Boys Club at RAND into their primary tool for the analysis of military strategy. By the early 1950s, both von Neumann and Morgenstern themselves were at RAND.

Von Neumann and Morgenstern developed their mathematical theory by analyzing the possible strategies and outcomes of two-player, zero-sum games. A zero-sum game is one in which one player's gain equals another player's loss. "Matching pennies" is a commonly cited example: Two players place pennies on a table at the same time. If both pennies are heads or both are tails, one player takes them; if they don't match, the other player takes them. Nash's contribution to game theory was to transcend the limitations of von Neumann and Morgenstern's simplified model by generalizing and validating their conclusions for multiplayer, non–zero sum games of strategy.

More ominously, Nash also influenced the social agenda of game theory by inventing a number of games of strategy to illustrate the potential usefulness of game theory for the social sciences. Nash's games, based on the classic Prisoner's Dilemma,[14] were explicitly noncooperative in nature, as opposed to games that encourage or at least permit cooperation among players.

The most notorious of Nash's games was provocatively named "Fuck You, Buddy." In order to publish it, it needed an alias that could be mentioned in polite company, so in print it became "So Long, Sucker."[15] (To avoid further repetitions of the coarse term here, the game will henceforth be referred to as "FYB.")

FYB begins with each of four players having seven chips. To win the game you must obtain all twenty-eight chips. You can only progress toward that goal by teamwork based on making agreements with other players. But to gain all of the chips, you will have to betray your teammates by reneging on what you agreed to. Strategic betrayal has become a familiar meme of popular culture thanks to "reality TV" shows like *Survivor*.

"When this game was tried out at dinner parties," a website for board game aficionados says, "a common outcome (reportedly) was that couples were so angered by the betrayals that they went home in separate taxis." One commenter added a warning: "Do not play with people who take things personally."[16] The antisocial attitude encapsulated in the game's title reflects a misanthropic view of human nature that is built into its rules: that all human behavior is motivated only by self-interest, and that rationality demands all players consider each other to be absolutely untrustworthy.

FYB established a pattern for games in which trickery, backstabbing, and blunt force are winning strategies, and trustworthiness is the currency of losers. That the game's principal author, John Nash, subsequently fell victim to a pathological condition characterized by irrational suspiciousness of others is not irrelevant. The point, however, is not to blame a mentally ill individual for the dangerous game of nuclear chicken with the Soviet Union. That responsibility lies with the civil and military officials who instilled paranoia into an entire society during the Cold War.

While (to my knowledge) none of Nash's collaborators at RAND were certifiably paranoid, they readily accepted his assumptions regarding human behavior. John von Neumann, *primus inter pares* among RAND's heavyweight intellectuals, believed that war was hardwired into human nature; his daughter spoke of "the deeply cynical and pessimistic core of his being."[17]

Mathematical Idolatry

Too many physicists and mathematicians who should know better allow themselves to believe that concrete knowledge of nature or society can be directly distilled from mathematics. That fatal philosophical fallacy—aptly designated "mathematical idolatry"—turns the relationship between science and mathematics on its head by making mathematics the master rather than the servant of science.

The RAND game theorists' version of the fallacy misapplied formal logic to real-life situations that lie far outside its scope. They erred at both ends of the process by starting with abstract mathematical postulates unmoored from space, time, and material reality, and ending up with mathematical models that model nothing that actually exists.

RAND's construction of nuclear military strategy on a foundation of game theory is perhaps the single most consequential example of mathematical malpractice. One-dimensional models that reduce human interactions to one-against-all antagonism produce inflexible strategies heavy on duplicity and blunt force, and light on intelligent efforts to resolve disagreements by diplomacy. By creating a framework within which US policymakers could only treat the USSR with unyielding hostility, RAND rationalized a Cold War strategy that humanity has thus far been fortunate to survive.

RAND and the Vietnam War

During the Eisenhower years, the uniformed military men had generally kept the RAND intellectuals at arms' length. After Kennedy took office in 1961, however, the RAND analysts rapidly gained influence among the military policy elite. Robert McNamara, Kennedy's choice to head the DoD, was a technocrat with a profound attachment to mathematical idolatry. He recruited a group of highly numerate intellectuals from RAND who came to be known as McNamara's Whiz Kids. Their mission was to develop a counterinsurgency strategy for Vietnam.

The RAND operatives went from scholarly studies of guerilla warfare to an ever-growing role in shaping Vietnam War policy. Former RAND analyst Benjamin Schwarz observes that the Whiz Kids "filled high posts in the Pentagon, where they planned the bombing strategies, assessed the pacification campaigns, and determined the budgets."[18]

"M&M": Mismeasuring Motivation and Morale

A signature RAND study, the Viet Cong Motivation and Morale Project ("M&M" for short), was devoted to investigating the guerilla fighters of South Vietnam to discover "what made them tick." It aimed at understanding why the Viet Cong[19]—the National Liberation Front, or NLF—so tenaciously resisted first French and then American colonization, which might hold the key to breaking that resistance.

The M&M project illustrates the intertwined relationship of RAND and DARPA and why it is often impossible to separate their contributions to the shaping of American science. The RAND social scientists worked out of DARPA offices in Saigon under contracts funded by DARPA. The researchers conducted extensive interviews with some twenty-four hundred NLF POWs, defectors, and refugees from battle zones. The studies consisted of two phases; the first, in 1964, was conducted by Joseph Zasloff and John C. Donnell, and the second, from January 1965 through 1967, was under the direction of Leon Gouré.

Zasloff and Donnell's team reported on "the idealism, commitment and high moral character which they found among the cadres of the NLF."[20] McNamara's people did not want to hear this, so RAND removed Zasloff and Donnell, and sent Gouré to take charge. That was the turning point from which RAND/DARPA social science in Vietnam devolved into pure opportunism, not reflecting objective research but providing made-to-order reports for Pentagon clients. Gouré devoted his efforts to discrediting Zasloff and Donnell's research and replacing it with false claims that NLF morale was in decline, and that the villagers who provided their social base were turning against them.

"The Answer Is Always Bombing"

When Gouré first arrived in Vietnam to take over the M&M study, in the taxi from the airport a RAND administrator casually asked how he intended to pursue the project. He patted his briefcase and said, "I've got the answer right here." Asked to elaborate, he replied, "When the Air Force is footing the bill, the answer is always bombing."[21] Gouré's reports dutifully stressed the centrality of the US bombing campaigns in demoralizing the enemy fighters. This analysis reinforced what RAND's air force clients—especially Gouré's personal friend Curtis LeMay—had most hoped to hear. LeMay, it will be recalled, had threatened to "bomb Vietnam into the Stone Age."[22]

McNamara and Lyndon B. Johnson were enthralled by Gouré's "light at the end of the tunnel" projections and encouraged their continuation. The destructive consequences of the bombing were of historically unprecedented scope:

> From March 1965 through November 1968, Operation Rolling Thunder unleashed 800 tons of munitions a day on North Vietnam, a total of a million bombs, rockets, and missiles. Even more bombs were dropped in the South with estimates ranging from seven million to eight million tons of them, not to mention 70 million liters of defoliants, as well as napalm and other anti-personnel weapons. Then, of course, there was the massive bombing of neighboring Laos and later Cambodia.[23]

And yet NLF "morale and motivation" did not collapse as Gouré continued to assure his clients it would.

There was a significant flaw in Gouré's scientific method. He was basing his conclusions on "an extremely selective use of the interview data."[24] Two of the social scientists working for him, Anthony Russo and Douglas Scott, were outraged when they discovered he had attached their names to a falsified report used to justify further bombing of rural Vietnam.[25] Russo raised a loud public protest.[26]

RAND itself, many decades after the fact, regretfully acknowledged its role in perpetuating the scientific deceit that justified the US slaughter in Southeast Asia. The think tank commissioned and published a book by one of its former M&M researchers, Duang Van Mai Elliott.[27] Mai Elliott's temperate narrative of ruinously misguided "intelligence gone wrong" is valuable for its frank revelations of blunders and errors, but acknowledges no criminal intent or criminal act on RAND's part. Anthony Russo, by contrast but not without reason, leveled a charge of "whitewash of genocide" against the M&M study and Gouré.[28]

RAND's encouragement of the CIA's Phoenix program of torture and systematic assassination in Vietnam merits particular notice. From the beginning of the Cold War, the CIA had spent many millions of dollars secretly developing new and better scientific psychological torture methods. "During the Vietnam War, when the CIA applied these techniques," historian Alfred McCoy writes, "the interrogation effort soon degenerated into the crude physical brutality of the Phoenix Program, producing 46,000 extrajudicial executions and little actionable intelligence."[29]

Anthony Russo, among other former RAND analysts, blew the whistle on their employer's criminal complicity in the Phoenix program. But Russo's numerous acts of courage were overshadowed by a much more significant one in which he assisted his colleague Daniel Ellsberg.

Daniel Ellsberg and the Pentagon Papers

Daniel Ellsberg was one of McNamara's Whiz Kids who has since become famous for a most un-RAND-like deed, but in a recent memoir Ellsberg tells us that in the beginning he was simply a typical, true-believer member of the Nuclear Boys Club.[30]

What brought about the change in Ellsberg's and Russo's views was a vast transformation that had occurred in the American public discourse during the second half of the 1960s. Deep revulsion over the Vietnam War had given rise to massive antiwar sentiment throughout American society, affecting every American institution, and the RAND Corporation was no exception.

Ellsberg's faith in RAND's Cold War mission was challenged, and finally destroyed, by the stark discrepancy between what he was personally witnessing as a researcher in Vietnam and what was being reported to the American public and the world. According to the official US military dispatches, the American forces were making solid, steady progress in winning an honorable war. In fact, they were being drawn ever deeper into a hopeless morass of war criminality. Finally, in 1971, Ellsberg made the fateful decision to reveal the truth in a most spectacular way.

Beginning in 1969, Ellsberg and Russo had surreptitiously photocopied thousands of pages of highly classified RAND documents that accurately depicted the political, diplomatic, and military history of the Vietnam War in meticulous detail. When the secret history was released to the *New York Times*, the *Washington Post*, and seventeen other newspapers, Ellsberg became the most consequential whistleblower in American history. Ellsberg's revelation will forever be known as the Pentagon Papers, but it would be more accurate to call them the RAND Papers. Among the deeply polarized public, Ellsberg was considered a hero by many and a traitor by many others. The Nixon administration went to great lengths to block the papers' publication and imprison the whistleblower, but those efforts proved unsuccessful and eventually self-defeating.

Ellsberg was arrested, indicted, and tried on espionage charges that could have landed him in prison with a 115-year sentence. The trial never went to verdict, however, and the charges were dropped "with prejudice," meaning that he could not be retried in the matter. Fortunately for Ellsberg, the Nixon administration's egregious use of "dirty tricks" against him had been exposed, causing the entire case to collapse.

Science and History—A Shining Example of Possibility and Actuality

Can social science—and history, in particular—really be practiced scientifically? Ironically, despite their dishonorable origins, the Pentagon Papers stand as a model of a genuinely scientific work of history. Robert

McNamara wanted to produce a scientifically accurate understanding of what had happened—and what had gone wrong—in Southeast Asia as a guide and caution to future policymakers.

To that end, McNamara mobilized a large team of first-rate RAND scholars, gave them full access to the Pentagon's files, and instructed them to create an *objective, unbiased, fully documented history* of the Vietnam War. What he absolutely did not want was for it to be read by the general public; it was intended to be "for official eyes only" for a long, long time. But among many other benefits to humanity, Ellsberg's exposure of the Pentagon Papers to the light of day revealed how a high degree of scientific objectivity can actually be attained in historical studies.

RAND "Diversifies"

After the Vietnam War, the RAND Corporation continued to exist as a think tank, but in a much altered form. Its official historians say that from the late 1970s onward, the policy institute "diversified" by serving clients other than Pentagon branches and performing research in fields other than military studies. A more accurate account would acknowledge that the Pentagon chiefs were so angry with RAND for its researchers' exposure of their secrets that the relationship was permanently damaged. The downgrading of the RAND Corporation changed the interface between science and the military, but it did not slow down the inexorable militarization of science.

*...And
the Only
Way
Out*

CHAPTER 22

Is a Science-for-Human-Needs Possible?

> Quand on est dans la merde jusqu'au cou, il ne reste plus
> qu'à chanter.
>
> —*Samuel Beckett*

SAMUEL BECKETT FAMOUSLY DESCRIBED the postmodern human condition as having us up to our necks in excrement. *All that's left for us to do now,* he said, *is sing.*

I disagree. In confronting the tragedy of American science, there is another option: replacing the current economic system with one not controlled by corporate power or addicted to military spending. That momentous transformation can prevail only on a global scale, but the road to accomplishing it runs through the United States.

Knowing what has to be done is easy. Getting it done—overcoming the formidable inertia of the status quo—is the hard part. Here are five essential planks in a platform to regain scientific integrity:

- 100 percent public *funding* of all scientific research and development; science 100 percent uncontaminated by corporate money.

- 100 percent public *control* of all scientific research and development.

- Vigorous, resolute governmental regulation of all scientific activities—*regulations with teeth.* Regulatory agencies such as the EPA, the FDA, and the CDC must be 100 percent independent of corporate influence and interference.

- Social costs must be accounted for in all cost-benefit analyses involving scientific policy and decision-making.

- The influence of the military-industrial complex must be exorcized. To paraphrase a sentiment expressed in a previous chapter:

  I don't want to abolish the military-industrial complex. I simply want to reduce it to the size where I can drag it into the bathroom and drown it in the bathtub.[1]

The struggle for uncorrupted science also points toward a society far more democratic than the United States has been since World War II. From the Hiroshima bombing to Operation Paperclip to universal NSA surveillance, the most significant decisions determining the course of American science have been made behind closed doors by mostly unelected policymakers. That is not what democracy looks like.

Logicians may point out that even if superseding capitalism is necessary to overcome the corruption of science, that does not guarantee that a socialist society would be fertile ground for valid science. Procapitalist ideologues have long insisted that socialism and science are as incompatible as oil and water.

Fortunately, this dispute needn't be confined to the realm of abstract speculation. There is a great deal of relevant historical evidence to turn to. Although history offers us no examples of fully developed socialism, a number of postcapitalist societies have existed. Their experience indicates that science motivated by human needs rather than private profit is not a utopian fantasy but a demonstrable reality.

Science and Revolution

The development of science from World War I to today was shaped by social revolution. The Russian, Chinese, and Cuban revolutions dynamited every facet of human experience out of their tradition-bound channels.

Throughout history, great social upheavals have created positive conditions for scientific advancement by removing obstacles to innovative thought and practice. In the process of "turning the world upside down," successful revolutions have typically eliminated censorship and broken the institutional power of entrenched intellectual elites that stifled science. Furthermore, by liberating subordinate social classes, revolutions have brought many more actors onto the stage of history. The resulting vast increase in the number of people able to play an active role in shaping their lives has enhanced all fields of human endeavor, including science.

The twentieth-century revolutions also encouraged the progress of science in other ways. The revolutions that replaced market-controlled economies with centrally planned economies created the capacity to marshal resources and focus attention on scientific goals to an unprecedented degree and with unprecedented results. "National liberation" revolutions in poorer countries broke the chains of imperialist domination that had previously restricted them to the low-tech role of raw-materials suppliers. Free at last to create their own modern industries, they turned to science and technology and made them foremost governmental priorities.

Science and the Russian Revolution

"The Bolsheviks who took over Russia in 1917," historian Loren Graham explains, "were enthusiastic about science and technology. Indeed, no group of governmental leaders in previous history ever placed science and technology in such a prominent place on their agenda."

In a period of sixty years, the Soviet Union made the transition from being a nation of minor significance in international science to being a great scientific center. By the 1960s, Russian was a more important scientific language than French or German, a dramatic change from a half century earlier.[2]

Despite the enthusiasm for science professed by Lenin and his colleagues, scientific development got off to a slow start in the early years of the Soviet Union. Efforts to promote research were severely hampered not only by the war-ravaged country's shortage of material resources but also by a deficiency of scientific talent caused by the exodus of many scientists who were hostile to the revolution. Nor did it help that a large proportion of the scientifically and technically trained specialists who did not emigrate were unsympathetic to the Bolshevik regime. More than a decade after the 1917 revolution, fewer than 2 percent of the Soviet Union's engineers—138 out of about 10,000—were members of the Communist Party.

Nonetheless, Lenin believed it would be counterproductive to try to forcibly impose the Bolshevik will on the recalcitrant scientists and engineers. Totalitarian control of scientific institutions was not his policy, but Stalin's. At the end of 1928 the Imperial Academy of Sciences, a Czarist institution, not only continued to exist but also was still the most prestigious of scientific bodies, and not one of its academicians were Communists. It was not until the period between 1929 and 1932, when Stalin was well on his way toward assuming complete command, that the Communist Party took over the academy and reorganized it.

From 1928 to 1931, Stalin promoted a cultural revolution (later to be imitated by Mao Zedong in China) that counterposed "proletarian science" to "bourgeois science." Purges of scientists and campaigns to ensure political conformity caused chaos and disruption within the scientific institutions. Scientific education was paralyzed, as the works of Einstein, Mendel, Freud, and others were condemned as bourgeois science and banned from the universities.

Meanwhile, however, the relentless pressure of external threats to the Soviet Union allowed Stalin to rally support, consolidate his power, and impose a program of rapid industrialization and agricultural collectivization requiring significant input from the sciences. Centralized planning plus massive funding rapidly gave birth to Big Science in the Soviet Union. The result was the creation of a powerful, but deformed, science establishment.

The limitations Stalin's policies imposed on free inquiry acted as a counterweight to the ability of the centralized economy to marshal and organize resources. The Soviet Union rose close to the top of the science world—second only to the United States. But in spite of its success in accomplishing impressive large-scale technological feats—hydroelectric power plants, nuclear weapons, earth-orbiting satellites, and the like— the achievements of the Soviet science establishment fell well short of its potential.

The demise of the Soviet Union in 1991 caused it to forfeit the strong position it had gained in international science. A 1998 assessment by the US National Science Foundation reported that with regard to Russia and the other former Soviet republics, science in those countries was on the brink of extinction, surviving only by means of charitable donations from abroad.[3]

Science and the Chinese Revolution

Just as World War I gave rise to the Russian Revolution, World War II likewise brought about a revolution in China under the aegis of a Communist Party. In 1949 the People's Republic of China was proclaimed, bringing to power a government that for the first time had the will and the ability to create institutions of Big Science, as had previously occurred in the Soviet Union.

Soviet science provided more than simply a model for Mao Zedong's regime. In the 1950s, Soviet scientists and technicians participated heavily in the construction of science in the new China, and they created it in their own image. There were, however, strings attached. Stalin expected the Chinese to submit to Soviet control, and that led to problems. Stalin had originally pledged full support to the effort to replicate Soviet Big Science in China, including the development of nuclear weapons. But there were sharp limits to the Kremlin's spirit of proletarian solidarity. When the Mao regime began to show signs of resistance to Soviet dictates, Soviet leaders apparently had second thoughts about creating a nuclear power in a large country with which it had a long common border. They reneged on their promise to share nuclear technology, precipitating a deep and bitter Sino-Soviet split.

In June 1960, Soviet premier Nikita Khrushchev abruptly ordered the withdrawal of all aid from China. Thousands of Soviet scientists and engineers were called home immediately, taking their blueprints and expertise with them. It was a ruthless act of sabotage that dealt a crushing blow not only to Chinese science but to the country's economic and industrial development as a whole.

Although it was set back several years, the goal of constructing a Soviet-style science establishment endured. The Soviet formula of heavily bureaucratized central planning plus massive funding produced similar mixed results in China. The country tested its first atomic bomb in 1964 and its first hydrogen bomb in 1967, and launched its first satellite into Earth orbit in 1970—number one in a series of scores of space probes leading up to 2003, when China became only the third nation to independently send an astronaut into space. The science establishment, however, remained focused on military and big industrial projects at the expense of research aimed at improving the lives of the billion-plus people of China.

It is undeniable that the centralization and planning made possible by the 1949 revolution are at the root of China's transformation from a negligible factor to a major player on the international science scene—perhaps even the primary future challenger to the United States' dominance. Yet the mass of the Chinese population continues to endure a material standard of living well below that of the people of Europe, Japan, and the United States.

A Necessary but Insufficient Step

The experience of the Soviet Union under Stalin and of China under Mao Zedong demonstrated that although replacing the market system with planned economy is a *necessary* step toward scientific redemption, it is certainly not *sufficient*.

The economies of the Soviet Union and China before the 1990s can accurately be characterized as *postcapitalist* but not *socialist*. The distinction is crucial. Fully developed socialism requires a high level of economic development and democratic control of production and political life. Both countries started out "*on the road* to socialism" but unfortunately got bogged down and eventually turned back.

Meanwhile, however, their impressive postcapitalist scientific achievements suggest that far more could be accomplished by societies meeting the criteria of genuine socialism. That a national science establishment more oriented toward human needs is possible has been demonstrated by a revolution that took place in a much smaller country.

The People-Oriented Science of the Cuban Revolution

In the first week of 1959, revolutionary forces under the banner of the July 26th Movement entered Havana and established a new government. As events unfolded, the revolution's leaders soon found themselves embroiled in conflict with their mighty neighbor to the north. They came to believe that economic sabotage by pro–United States industrialists operating within Cuba could only be prevented by nationalizing the Cuban economy and declaring a governmental monopoly of foreign trade. The nationalization of firms owned by US investors provoked a harsh confrontation that has persisted to the present day. For protection, the new Cuban government entered into an alliance with the Soviet Union.

Once the revolution's leaders were in command of a fully national-ized economy, they enjoyed the same advantages that had enabled their Soviet and Chinese counterparts to develop powerful science establish-ments. The situation in Cuba, however, was considerably different.

The earlier revolutions had occurred in two of the world's largest countries, but Cuba was a small island of fewer than ten million in-habitants. Its scientific endeavors, therefore, were not channeled into a quixotic effort to compete directly with the United States in the field of military technology. Instead, Cuba would depend on diplomatic and po-litical means for its national security—that is, on its alliance with the So-viet Union and on the moral authority the Cuban revolution had earned throughout Latin America and the rest of the world. That allowed its sci-ence establishment to direct its attention toward nonmilitary pursuits.

The Cuban leaders oriented their science program toward the solu-tion of social problems. Scientific development, they decided, depended first of all on raising the educational level of the entire population. Before the revolution, almost 40 percent of the Cuban people were illiterate. In 1961 a major literacy campaign was launched that reportedly resulted in more than a million Cubans learning to read and write within a single year. Today the literacy rate is 99.8 percent[4] and science education is a funda-mental part of the national curriculum.

In addition to education, universal health care was assigned high priority, giving impetus to the development of the medical sciences. A punitive economic embargo imposed by the United States compelled the Cubans to find ways to produce their own medicines. They met the challenge, and the upshot was that Cuba, despite its relatively humble economic status, now stands at the forefront of international biochemi-cal and pharmacological research.

Cuban Medical Practice

As evidence of the success of their medical programs, Cuban officials point to comparative statistics routinely used to quantify the well-being of nations, the most informative measures being average life expectancy and infant mortality. In both categories, Cuba has risen to rank among the wealthiest industrialized countries. Richard Levins of Harvard Universi-ty's School of Public Health observed that "Cuba has the best healthcare in the developing world and is even ahead of the United States in some areas such as reducing infant mortality."[5]

Another key indicator of the quality of a nation's health care system is the doctor-to-patient ratio. According to World Health Organization statistics, in that category Cuba ranks number one out of 194. There are 7.92 doctors for every thousand people in Cuba, compared with 2.55 doctors per thousand in the United States, 2.81 in the United Kingdom, and 3.23 in France. Most of the nations of the Third World have fewer than one doctor per thousand inhabitants.[6]

Cuban "Medical Internationalism"

The doctor-to-patient ratio tells only part of the story, because Cuba's medical schools produce a large surplus of physicians—far more than can be put to productive use on the island itself. As a result, Cuba has actively exported doctors to other parts of the globe. The itinerant Cuban physicians do not "follow the money"—they go to the countries most in need of health care services.[7] With the stated ambition of becoming a world medical power, Cuba offers more humanitarian medical aid to the rest of the world than does any other country, including the wealthy industrialized nations. In addition to the medical equipment, medicines, and the services of some fifty thousand doctors working in sixty-seven countries, Cuba has also helped to build and staff medical schools in Ethiopia, Guinea-Bissau, Yemen, Guyana, Uganda, Gambia, Ghana, Equitorial Guinea, and Haiti.[8]

The Trump administration abruptly ended a brief period of US "normalization of relations" with Cuba that had begun in late 2014, and Cuba's medical internationalism has been a particular target of its hostility. "Thousands of Cuban medical personnel have been withdrawn from Brazil, Ecuador and Bolivia in 2019" due to US pressure on Latin American governments, inflicting heavy damage on public health programs in those countries.[9]

Cuban Medical Science

Cuba's health care successes have been closely linked to pioneering advances its laboratories have produced in the medical sciences. In the 1980s a worldwide biotechnology revolution occurred, and Cuban research institutions took a leading role. Cuban bioscience produced the first effective vaccine against meningitis B and the first human vaccine containing a synthetic antigen (the active ingredient of a vaccine).[10] Other noteworthy products include the popular cholesterol-reducer PPG (which is derived from sugarcane), monoclonal antibodies used to combat the rejection of transplanted organs, recombinant interferon products for use against viral infections, epidermal growth factor to promote tissue healing in burn victims, and recombinant streptokinase for treating heart attacks.

The Cuban biotech institutes focus their attention on deadly diseases that the profit-motivated multinational drug corporations tend to ignore because they mainly afflict poor people in the Third World. An important part of their mission is the creation of low-cost alternative drugs.

The synthetic antigen vaccine mentioned above was developed to treat Hib (Haemophilus influenzae type b), a bacterial disease that causes meningitis and pneumonia in young children and kills an estimated four hundred thousand throughout the world every year.[11] An effective vaccine against Hib already existed and had proven successful in industrialized nations, but its high cost sharply limited its availability in the less affluent parts of the world.

What Cuban Science Can Teach the Rest of the World

The Cuban example offers a particularly clear case study of how a social revolution has affected the development of science. The revolution removed the greatest of all obstacles to scientific advance by freeing the island from economic subordination to the industrialized world.

The Cuban scientific achievements testify that important, high-level scientific work can be performed without being driven by the profit motive. They also show that centralized planning does not necessarily have to follow the ultrabureaucratized model offered by the Soviet Union and China, where science served the interests of strengthening the state instead of improving the well-being of their populations.

Revolutionary Cuba has come closest to realizing the goal of a fully human-oriented science. Although Cuba's small size limits its usefulness as a basis for universal conclusions, its achievements in the medical sciences certainly provide hope for the rest of the world. If Cuban science has done so much with so little, what marvels could the mighty engine of American science achieve?

Restoring American Science and Society to Health

The multidimensional crisis of American capitalism will no doubt continue to stimulate the renaissance of socialist politics in the United States. The aspiration of socialist politics, however, should not be limited to electing politicians who call themselves socialists, but to building a constituency, a political base, a *social movement*, with the numerical power and the determination to make fundamental changes in the economic system. Those changes include, for starters,

nationalizing Big Pharma,

nationalizing the fossil fuel and electrical power generation industries,

nationalizing the banks and insurance companies,

and nationalizing the swordmakers and retooling them for plowshare production.

Some of these ideas began to edge their way into the national discourse in the 2020 election campaign.[12] Nationalizing key industries is a crucial step toward creating *economic democracy*, which will require a monumental struggle to abolish private ownership of the country's industrial base and other productive resources.

The idea most essential to any significant social progress, however, has thus far gained little traction: the need to challenge and break the chokehold of militarism on US policy. Until a sizable segment of the general public understands that American GIs are not heroes "defending our freedoms" but victims of a brutal war machine serving US corporate interests, American science will remain chained to the chariot of destruction, and "the dream of lasting peace," as Bob Marley sang, "will remain but a fleeting illusion."[13]

Scientists and tech workers have a vital role to play in constructing a better world, but they cannot prevail on their own. Although antiwar, antinuclear, environmental, racial justice, and other protest movements have not yet proved capable of forcing Big Science onto the path of social responsibility, they represent the nuclei of a massive movement for social change. They need to continue to grow, to coordinate their efforts, to strike together ever harder and more often. Most necessary is the resurrection of a powerful American labor movement that can embrace and unite them all.

The ultimate problem to be solved is whether science, technology, and industry can be brought under genuinely democratic control in the context of a global planned economy, so that we can communally put our hard-won scientific knowledge to mutually beneficial use. We're a long way from there now. Samuel Beckett had a point: We are up to our necks in *merde*. But rather than singing, we'd better get busy organizing a colossal collective effort to dig ourselves out.

There is one powerful reason to believe we can ultimately succeed. I urge you to never forget that American policy today is controlled by a remarkably small number of people. For every one of them, there are thousands of us. *We are many; they are few.* And therein lies our hope for the future.

EPILOGUE

The COVID-19 Pandemic

THE ROOT OF THE WORD "history" is "story." My job as a historian of science in this instance is to integrate the story of the COVID-19 pandemic into the larger story of how science has developed in the United States since World War II.

The science most essential to the immediate story is epidemiology, the study of how often diseases occur in different populations, and why.[1] In times of crisis, epidemiologists are called away from their research posts and marched before television cameras to predict how widely the menacing disease agents might spread and how deadly they might prove to be. Like weather forecasters, they cannot project a precise picture of future developments, but the expertise they bring to the task is invaluable. Then their colleagues in virology and immunology are tasked with the urgent development of diagnostic tests, treatments, and vaccines to combat the dangerous pathogens.

The Timeline of the Initial Outbreak

That a previously unknown virus posed a potential threat to humanity of extraordinary proportions had become evident by New Year's Day 2020, following the outbreak of a devastating illness in Wuhan, China. The Chinese authorities, hoping the contagion would soon recede, originally tried to block reports of its existence, but the attempt at censorship quickly collapsed. By the last week of January, Wuhan, a city of 11 million people, had been placed under lockdown.

279

Public health efforts to isolate the diseased population and prevent spread of the virus proved futile. The timeline of contagion advanced rapidly. On January 30, the World Health Organization declared a global health emergency. Within three weeks, the virus appeared in other parts of Asia, in Europe, in Latin America, and in Africa, and on February 28 it claimed its first fatality in the United States. Two weeks later the WHO officially declared a pandemic. Although researchers could not specify the rates of infection and mortality with precision, early indications suggested they could be high enough to infect billions of people and kill tens of millions. Epidemiologists warned that a worst-case scenario could bear comparison with the "Spanish flu" pandemic of 1918,[2] in which a third of the world's population was infected and fifty million people died worldwide.[3]

The social disruptions in response to this possibility were by necessity swift and severe. On March 9, the Italian government placed Italy's entire population of 60 million people under lockdown. Schools and universities were closed, public events were banned, and commercial and retail businesses were forced to cease operations. The impact on Italy's national economy was devastating, and as similar measures were adopted throughout the world, the global economy contracted accordingly.

Although the Trump administration's response to the crisis was characteristically fatuous and obstructionist—first denying, then dithering—nongovernmental institutions in the United States took the lead by shutting down their own operations. Some of the most iconic elements of American popular culture suddenly vanished, as Major League Baseball and the National Basketball Association suspended their schedules and Broadway theaters went dark. Airlines and cruise ships canceled flights and voyages, restaurants and bars shut their doors, and businesses large and small sent their employees home, causing millions of workers to fear the loss of their livelihoods. With the world's two largest economies—that of the US and China—rapidly seizing up, the global economy threatened to go into free fall.

In an effort to slow the spread of the disease, public health officials introduced some new phrases into the public discourse: "social distancing" and "sheltering in place." Over the course of a few weeks, state by state, Americans were advised to stay at home as much as possible; to avoid all gatherings of more than fifty people, then ten; and to stand no closer than six feet to anyone else. Businesses deemed nonessential were ordered closed in many places. Jobs were restructured, when feasible, to allow workers to work from home. Schools of all kinds, from universities to neighborhood yoga studios, retooled their operations to provide all instruction online. Regardless of how the story of this particular viral pandemic develops, it has given us a harsh glimpse of a dystopian, "socially distanced" future. At the same time, it has reminded us of the value of human contact and challenged us to maintain solidarity in the face of isolation.

Science and Public Policy

In the realm of science, epidemiology and its allied disciplines have come to the fore in the current crisis. At the beginning of April, the *New York Times* offered this assessment:

> While political leaders have locked their borders, scientists have been shattering theirs, creating a global collaboration unlike any in history. Never before, researchers say, have so many experts in so many countries focused simultaneously on a single topic and with such urgency.[4]

Chinese virologists, microbiologists, and other researchers rapidly deciphered the novel coronavirus's genetic code—an indispensable step toward the creation of a vaccine. Much to their credit, they did not hoard the vital knowledge as "intellectual property" but immediately published it for every researcher in the world to see and to use. If they had not announced their findings, the virus's gene sequence would eventually have been worked out elsewhere, but the delay would surely have contributed to more deadly consequences.

Big Pharma, meanwhile, has resisted the advance of scientific internationalism. Drug firms developing experimental treatments have come under intense investor pressure "to consider ways they can profit from the crisis."[5] American drug companies in particular have powerful friends in high places. Alex Azar, Trump's Health and Human Services secretary, is a former head of pharma giant Eli Lilly's US division and board member of the drug lobby group Biotechnology Innovation Organization. When asked about the affordability of a prospective coronavirus vaccine during congressional hearings in late February, Azar responded, "We can't control that price."[6]

The good work of scientists did not translate into beneficial social policy. The Chinese government's initial attempt to deny and cover up the viral outbreak in Wuhan cost the world weeks of valuable time to respond to the crisis. Eventually, however, the prospect of economic collapse forced the regime to recognize the seriousness of the outbreak, and it began to take action to contain it. Although its compulsory quarantining efforts proved effective, they came too late to prevent the spread of the virus beyond Chinese borders.

In the United States, the Trump administration's initial response to the pandemic was egregiously irresponsible. In mid-March, Dr. Ashish Jha, director of the Harvard Global Health Institute, declared:

> By every objective measure, the American response has been dramatically worse than everybody else's. Here we are, two months after the WHO test kits became widely available, and we still don't have widespread availability of testing in America. We've had two months of mixed messaging from the government, saying "This is not a big deal." . . . That's one misstep after another after another after another. It's been an abysmal failure.[7]

Trump began by dismissing the threat as "fake news" and as a hoax perpetrated by his Democratic Party rivals. When the country began shutting down around him, he shifted to pretending he'd never doubted the seriousness of the situation and tried to excuse his inaction by exclaiming that the coronavirus "came out of nowhere"[8] and "blindsided the world."[9]

The Assault on Science That Put Us All at Risk

If Trump and his associates had indeed been taken by surprise, they had themselves to blame. In May 2018 his administration disbanded a National Security Council directorate at the White House that had been set up precisely to warn of possible pandemic outbreaks. A former director of the unit stated that shortly before Trump took office, the group had been monitoring "a rising number of cases in China of a deadly strain of the flu and a yellow fever outbreak in Angola."[10] Then, in September 2019, just two months before the earliest warning signals emerged from Wuhan, the administration closed down a USAID early-warning program named PREDICT. Its researchers had identified 1,200 viruses, including 160 novel coronaviruses "that had the potential to erupt into pandemics," and had trained scientists in 60 foreign laboratories, including the one in Wuhan that first sounded the alarm.[11]

Trump's first move toward coordinating a response to the crisis evinced his notorious disdain for science, scientists, and expertise in general. On February 26, he appointed vice president Mike Pence to head up the effort—a man notorious for declaring, "Global warming is a myth," asserting that "smoking doesn't kill," and promoting "intelligent design theory" in opposition to evolutionary biology.[12] Pence's appointment was a natural extension of the administration's three-year war on science, which in turn was the fruit of decades of corporate campaigns against science devised by right-wing ideologues.

The assault on science significantly weakened the federal regulatory agencies charged with oversight of scientific affairs by slashing the budgets of the CDC, the EPA, the FDA, and others. A spokesperson for the Union of Concerned Scientists charged that "the administration's relentless dismantling of our overall federal public health science infrastructure" had driven out hundreds of "dedicated infectious disease experts, epidemiologists, triage managers, and heads of public university research labs" and replaced them with industry lobbyists and political ideologues.[13]

Trump's first two years in office saw more than 1,600 scientists leave government service. The EPA's staff decreased by a third. In 2018 it was revealed that the CDC was "dramatically downsizing its epidemic prevention activities" around the world by 80 percent.[14] Among the administration's first acts in response to the coronavirus outbreak was to propose reducing CDC funding by 16 percent and cutting spending for global health programs by $3 billion. Should anyone be surprised that the debilitated CDC bungled its first responsibility in defending against the new coronavirus—diagnostic testing?[15]

The Testing Fiasco

The agenda for medical science is summed up by the triad *diagnosis, prognosis, therapy*. In the case of a new viral infection, the first two would require a test to determine the speed, geographic range, and virulence of its spread, and the third calls for an effective antiviral treatment for infected individuals, as well as the development of a vaccine to prevent future infections.

A few days after Chinese virologists published the COVID-19 genome, not only had German researchers already developed a diagnostic test for the disease, but a small German company had also begun mass-producing and distributing it. This fifty-five-employee outfit was able to produce a million high-quality tests a week. By the end of February, it had provided the World Health Organization with 1.4 million tests, which the WHO in turn distributed to sixty countries throughout the world.[16] The United States was not among them, and administration officials offered no explanation for the costly missed opportunity.

Meanwhile, soon after the appearance of the German test, the CDC announced the creation of its own test and shipped 160,000 kits to laboratories throughout the United States. Lamentably, the CDC tests proved to be flawed and unusable. Testing efforts in the United States came to a standstill, leaving health officials in the dark regarding the geographic distribution of the infection and the speed of its diffusion. The inexcusable delay in testing heightened the pandemic's threat to the American public.

Beyond the Emergency:
Yet Another Anthropogenic Existential Threat

If COVID-19 does not decimate the world population in its first appearance, the current emergency will eventually ease and a natural yearning for a return to normalcy will surface. Quarantines will be canceled, the economy may take tentative steps forward, and epidemiologists will be sent back to their research stations. But if the contagion's wake-up call is not heeded and complacency prevails, an invaluable opportunity will have been squandered.

The history of the 1918 influenza pandemic stands as a powerful warning against any such complacency. It attacked in several waves. The first wave, which struck in early March 1918 and subsided a few months later, was relatively mild. But the second and third waves, in the autumn of 1918 and the following winter, were far more lethal, accounting for most of the tens of millions of fatalities worldwide.

The 1918 pathogen has been identified as an avian virus—a bird flu—now designated as H1N1. Epidemiologists believe that the strain that first appeared in March 1918 then mutated into a more virulent strain, which caused the great devastation of the second and third waves. Not alarmism but prudence compels us to recognize that the novel coronavirus of 2020 could follow a similar path. In April, the director of the CDC warned that "the assault of the virus on our nation next winter" is likely to be even more devastating.[17]

Furthermore, beyond COVID-19, there are solid reasons to fear that the generation of dangerous new pathogens can no longer be considered a once-in-a-century occurrence. The conditions that allow them to emerge continue to exist. Modern agribusiness practices have created disease-riddled "influenza factories" that can be expected to churn out novel viruses with increasing regularity.[18] The public discourse had already recognized two existential threats to the human race—thermonuclear war and climate change. The danger of a "perfect storm" global contagion adds a third. As a byproduct of agribusiness, it, too, is anthropogenic.

The Perfect Storm

The virus that causes COVID-19 is a single strand of an elongated nucleic acid molecule enclosed in a protein coating. Viruses come in either DNA (deoxyribonucleic acid) or RNA (ribonucleic acid) varieties, but those that cause the biggest problems for humans are of the RNA type. RNA viruses, lacking the error-correcting sophistication of DNA replication, have the highest mutation rates of any organisms on Earth.[19] Some of those mutations enhance the ability of viruses to leap from birds, pigs, and other animals to humans. Other mutations have made some viruses more contagious, and others more deadly. Yet another kind of mutation can result in a virus with a long incubation period, during which people with no disease symptoms can nonetheless transmit it to others.

The perfect storm of a deadly viral strain would be one that maximized all four of those traits: it could infect humans, hide itself for a while, jump easily from victim to victim, and kill a large proportion of those infected. The odds of such a virus emerging in nature are infinitesimally small, but modern industrial agriculture has created conditions that—to use a familiar phrase—now make it a matter of *not if but when*.

First reports in early 2020 of the novel coronavirus in Wuhan gave epidemiologists reason to fear that this could well be the nightmare they had anticipated. One of those epidemiologists, Rob Wallace, had declared four years earlier that "southern China has served, and will serve, as ground zero for influenza panics." Furthermore, "much of the world's economic productivity," he warned, "stands to suffer catastrophically if a deadly pandemic were to erupt, for instance, in southern China."[20] Wallace was not peering into a crystal ball; he was demonstrating the predictive power of a science heretofore underappreciated by the general public and policymakers alike.

Revenge of the Poultry

If you are an informed citizen, chances are you're aware of the moral depravity of modern egg production. Here is a brief, but sadly accurate, description of the extreme cruelty it entails:

> Virtually all chickens start off life at industrial hatcheries. Male chicks, who will never lay eggs and don't grow fast enough for meat production, are typically ground up alive. Females have the tips of their sensitive beaks cut or burned off before they are shipped to egg factory farms. This painful mutilation is meant to prevent hens from pecking each other in stressful, overcrowded conditions.
>
> At most egg factory farms, hens are crammed into barren wire cages. These cages are stacked in tiers inside giant window-less sheds that can run the length of two football fields. Each bird has less floor space than the size of a sheet of notebook paper for nearly her entire life. In such extreme confinement, these poor animals can't even spread their wings, much less move around without stepping on and climbing over other hens.[21]

The hens' lifespan is eight to ten years, but they are usually slaughtered after twelve to eighteen months when their egg production drops. Well-meaning consumers try to counter these practices by paying high prices, if they can afford to, for eggs allegedly laid by "free-range chickens."[22] Some states have attempted to ban the abhorrent factory-farming practices, or at least regulate them. And yet industrial egg production continues to grow, both in volume and in geographical scope.

Gigantic factory farms of this type first arose in the American South after World War II, but the drive to maximize profit led multinational poultry and egg producers such as Tyson Foods and the CP Group[23] to replicate them overseas, in poorer countries where even inadequate US regulation would not apply and lower wages could be paid. Their spread was relentless, as this comparison reveals:

> 1929: The total chicken population in the US was 300 million, with an **average flock** size of **70 birds**.

> 1992: The total global poultry population was 6 billion, with an **average flock** size of **30,000 birds**. Ten years later, in 2002, the total was 9 billion.[24]

The aforementioned CP Group, one of the world's largest conglomerates, "operates joint-venture poultry facilities across China, producing 600 million of China's 2.2 billion chickens annually sold."[25]

Although it brings no relief to the victimized birds, they are silently avenging themselves for the suffering inflicted upon them by spawning plagues to torment the human race. Crammed together by the tens of thousands in warehouses throughout the globe, the birds' unhealthy bodies create a huge pool of opportunities for viruses to come forth, multiply, and mutate.

Poultry farms are the most notorious incubators of viral disease ("bird flus"), but they are not the only culprits. Others are hog farms ("swine flus") and, increasingly, the incorporation of wild animals, including bats, into agribusiness operations.[26] And because these fertile breeding grounds are often located in close proximity to population centers, the transmission of their viruses to humans is virtually assured.

Be advised: despite what grocery store labels claim, the farmed animals are not "natural." The chickens, ducks, and geese on the poultry farms, for example, are monocultures—domesticated birds that have been selectively bred to possess a high degree of uniformity. Their resulting lack of diversity makes the nearly genetically identical fowl all the more susceptible to raging contagion. At the first signs of illness, the factory farms pump their flocks full of antibiotic and antiviral drugs, prompting natural selection to create "superbug" pathogens with ever-higher resistance to the pharmaceuticals.

The preponderance of evidence suggests that the mutant coronavirus causing COVID-19 first appeared in bats. This in no way absolves capitalist agribusiness of responsibility. The chain of causation is complex, and whether wild or farmed bats were the virus's direct source, Big Food indisputably created the underlying social and ecological conditions for the pandemic's emergence. "As industrial production encroaches on the last of the forest," novel pathogens "find their way onto a truck, whether in food animals or the labor tending them." That is how a lethal virus "can suddenly find itself spilling over into humans in the big city only a few days out of its bat cave." Furthermore, "the greater the extent of adjunct deforestation, the more diverse (and exotic) the zoonotic pathogens that enter the food chain."[27]

The bottom line is that industrial poultry and livestock farms pose a clear and present danger to humans, and emergency responses every time a new killer virus appears are not an adequate defense. "Clearly humanity shouldn't *start* reacting to a pandemic when it's already underway," Rob Wallace declares. "Let's stop the outbreaks we can't handle from emerging in the first place."[28] Everyone—even those who have never been moved to protest the appalling treatment of our feathered friends—should recognize that it is in every human being's self-interest to demand an end to the factory farms.

Trump Is But a Symptom of a Systemic Failure

Understanding the COVID-19 pandemic not as a unique emergency but as one episode in a series of growing threats attests that the Trump administration, for all its perfidious bungling, is not the main story. Trump and his allies made a terrible situation worse but did not cause it, and voting them out of office will not make the problem go away. Much more profound remedies are necessary, and that message seems to be gaining unprecedented attention from the American public.

The exposure in 2020 of inadequate medical supplies and hospital capacity throughout the United States disclosed a crisis four decades in the making. The shortage of intensive care unit beds and ventilators is not a matter of poor planning or criminal negligence; it was a crime committed with intent. It was the culmination of a policy of *intentional downsizing of hospital emergency capacity* that began during the Reagan administration and has continued to the present.

The shortage of hospital beds and medical supplies is not a case of criminal negligence. It was a crime committed with intent.

Historian Mike Davis describes the consequences of "years of profit -driven cutbacks of in-patient capacity":

> According to the American Hospital Association, the number of in-patient hospital beds declined by an extraordinary 39 percent between 1981 and 1999. The purpose was to raise profits by increasing "census" (the number of occupied beds). But management's goal of 90 percent occupancy meant that hospitals no longer had the capacity to absorb patient influx during epidemics and medical emergencies.[29]

The profit-driven cutbacks of in-patient capacity persisted into the twenty-first century, through Republican and Democratic administrations alike. Serious hospital bed shortages in both the 2009 and 2018 flu seasons foretold danger ahead, but policymakers studiously ignored the warnings. By March 2020, US hospitals could provide only 65,000 intensive care unit beds, while "government planning assumptions based on past flu pandemics suggest a surge in demand for intensive care that could range somewhere between 200,000 and 2.9 million patients."[30]

Davis also exposed the roots of another critical deficiency in pandemic preparedness: Big Pharma's underinvestment in antibiotic and antiviral research because it offers less profit potential:

> Of the eighteen largest pharmaceutical companies, fifteen have totally abandoned the field. Heart medicines, addictive tranquilizers, and treatments for male impotence are profit leaders, not the defenses against hospital infections, emergent diseases, and traditional tropical killers. A universal vaccine for influenza—that is to say, a vaccine that targets the immutable parts of the virus's surface proteins—*has been a possibility for decades* but never profitable enough to be a priority.[31]

The Macroeconomic Fallout

This public health crisis has triggered the fastest economic meltdown in American history, exposing the inadequacy of the US economic system to cope with a challenge of these dimensions. With the public interest under siege, the private enterprise system flailed and failed. As the country was confronted with massive job losses and a rapid decline in economic activity, wealthy investors scrambled to protect their own business interests, and all hopes for salvation focused on a $2.2 trillion federal bailout, which quickly proved to be insufficient. The limitations and unreliability of the current economic system have no doubt only begun to be exposed.

In the first days of 2020, I wrote in the concluding chapter of this book that the tragedy of American science could not be resolved without a total overhaul of the country's economic system, and proposed as necessary steps in that direction the nationalization of Big Pharma, the fossil fuel industry, the banks, the insurance companies, and the military contractors. Nationalization alone is not enough, but it can lead toward *socialization*—a nationalized economy administered by a government that represents the entire society rather than a minuscule superwealthy elite.

I cited, in a footnote, some harbingers of hope that demands for nationalizations were just beginning to gain serious attention in the public discourse. But in the wake of the economic chaos triggered by the pandemic, what began as a trickle has become a flood, as these headlines attest:

> Will COVID-19 Force Nationalization of Manufacturing?[32]
>
> COVID-19 Crisis Fighters Say the US Must Nationalize Healthcare Equipment Production[33]
>
> It's Time to Nationalize the Airlines[34]
>
> Green Group Urges Nationalization of Oil and Gas Industry Amid Coronavirus Outbreak and Economic Upheaval[35]
>
> France, Italy, Spain in Bid to Nationalize Businesses, Hospitals as Virus Hits[36]

Pillars of establishment journalism in the United States began publishing opinion pieces calling for federal intervention in the US economy. A *New York Times* article headlined "How the World's Richest Country Ran Out of a 75-Cent Face Mask" attributed the scandalous lack of pandemic readiness to "capitalism consuming our national preparedness and resiliency." The author lamented:

> What a small, shameful way for a strong nation to falter: For want of a 75-cent face mask, the kingdom was lost. . . . Perhaps the only way to address the shortfall now is to recognize that the market is broken, and to have the government step in.[37]

The Trump administration, not surprisingly, angrily rejected calls for nationalizations, but their fury revealed that the enemies of science and reason are now on the defensive. Meanwhile, they found themselves obliged to invoke the Defense Production Act to produce ventilators and other critical medical supplies, thus demonstrating that the investor class is not essential to the production system. At the same time, the need for a multitrillion-dollar federal intervention dealt a powerful rebuke to right-wing "small government" demagogy. These remarkable turnabouts in the national conversation should encourage all who recognize the pivotal nature of the times we live in.

Socialism or Dystopia?

Will the social dislocations provoked by COVID-19 endure for weeks, months, or years? If it goes away, will it come back? If so, when and how often? If not, will more deadly viruses arise every flu season? These are questions not even the epidemiologists can answer.

These unknowable aspects of our future make our present uneasy. Will the tens of millions of jobs that vanished in an eyeblink ever reappear? Will visitations of these plagues periodically condemn us to go stir-crazy in self-quarantine? To don gloves and masks every time we go out the door? To forever "socially distance" ourselves from friends, neighbors, and loved ones? To endure a dismal existence without sports, concerts, plays, parties, and restaurants, and with only "virtual" museums?

But one important fact is known. As long as our collective fate is in the grasp of an economic system that allows industrial poultry and livestock farms to generate novel viruses, the constant threat of recurring dystopia will hang over our heads.

As long as our collective fate is in the grasp of an economic system that allows industrial poultry and livestock farms to generate novel viruses, the constant threat of recurring dystopia will hang over our heads.

The agribusiness proprietors of the factory farms are aware that dangerous viruses are their inevitable byproducts. They are also aware that the social costs of the pandemics they create are not borne by them. They expect capitalist governments to keep cleaning up the deadly messes their infections cause. The early indications of the 2020 pandemic, however, raise doubts as to whether capitalist governments are capable of doing so. Multitrillion-dollar corporate bailouts may prove inadequate to save the world capitalist order from total meltdown. And then what?

The stakes have been raised to the sky. The need to replace the global economic system that serves private interests with one that serves the public interest is more urgent than ever. In the course of that struggle, American science, from epidemiology to etiology to ecology to economics, can redeem itself by socializing its practice and illuminating the path forward.

A thorough socialization of the American economy, combined with democratic determination of national science policy, would free American science from corporate bondage. By replacing business competition with cooperation as the driving force of science, the patent system could be rendered irrelevant, and secrecy could give way to knowledge sharing. Social ownership of Big Pharma and for-profit health care organizations could shift research to federal laboratories to create vaccines to protect the public, not enrich investors. Social ownership of Big Food could begin the transition to regenerative agriculture, with technologies geared not to profiting food corporations but to revitalizing the soil and preventing the industrialized production of viral pandemics. If we can survive the crisis at hand, resolving the tragedy of American science is not a hopeless cause.

ACKNOWLEDGMENTS

I THANK MICHAEL Steven Smith, Debby Smith, and the late Frances Goldin for inviting me to contribute to a collection of essays, *Imagine: Living in a Socialist USA*, which eventually led to my writing this book.

I also owe Frances, founder of one of America's premier literary agencies, appreciation for opening the doors to publication for me. Her mission, she always said, was getting "books that make a difference" published; I am deeply gratified that she deemed this one to be in that category. More specifically, without the talent and efforts of Sam Stoloff, one of Frances's colleagues, this book would likely not have escaped the confines of my hard drive.

I am likewise grateful to Nisha Bolsey, Rory Fanning, and Anthony Arnove of my favorite publishing house, Haymarket Books, for their encouragement and support. Another key person in the creation of this book was Christopher Dols, who recommended its publication to Haymarket. Chris is one of the prime movers behind the rebirth of *Science for the People*, an activist organization essential to any hope of resolving the tragedy of American science.

My gratitude to Ruth Baldwin is boundless for all the hard and caring work she put into editing and improving this book. Ruth had previously edited my *People's History of Science*, and when Haymarket Books told me she was available to collaborate with me on this one, I was elated at the prospect of working with her again. I was further blessed, in the later stages of the book's production, with the editorial collaboration of Caroline Luft, who gave the text a thoughtful, critical reading to assure that every sentence expressed exactly what I wanted it to say.

I have already acknowledged, in the dedication, the indispensable contributions of investigative journalists, whistleblowers, and others who allow us an occasional peek behind the curtains of corporate and governmental secrecy. Some of their names appear in the text, and others in the endnotes.

A number of friends and colleagues with specific areas of expertise have collaborated in this project by reading and critiquing early drafts of chapters, by helping me gain access to key sources, or by helping me work through difficulties of interpretation. Listing their names here hardly suffices to express how valuable their input was to me, and how much I appreciated it, but here goes anyway: *Grazie mille* Carol Dansereau, Richie Feinberg, Sid Finehirsh, Mark Friedman, Betsy Samuelson Greer, Lynn Henderson, Marc Henley, Phyllis Kittler, Paul LeBlanc, Guy Miller, Fred Murphy, Michael Schreiber, Barry Sheppard, Ernie Tate, and Tom Tilitz. I thank Sheldon Krimsky and Rob Wallace for their wise counsel. Professor Krimsky's suggestions, based on his extensive study of the many ways science is corrupted, were invaluable, and Rob Wallace's critique of the epilogue was especially appreciated for helping me avoid epidemiological faux pas.

A special shout-out to Dennis Edge: before I had written a single line of this book, I described the concept to him, and he created a poster design that I framed and put on the wall in front of my desk. It continuously provided me with inspiration to write a book worthy of his powerful illustration.

As it was in the beginning and will be in the end, for me the font of inspiration that surpasseth all science is my cherished companion, *mi vidita Maruja.*

ENDNOTES

New York Times will be abbreviated "NYT" throughout the notes.

Preface to the 2022 Edition

1. Nitasha Tiku and Jay Greene, "The Billionaire Boom," *Washington Post*, March 12, 2021.
2. CBS Sunday Morning, January 23, 2022.
3. Oxfam, "Inequality Kills," January 2022.
4. Meanwhile, COVID-19 deaths worldwide surpassed six million. "Number of Novel Coronavirus (COVID-19) Deaths Worldwide as of March 28, 2022, by Country," statista.com.
5. Sydney Lupkin, "How Operation Warp Speed's Big Vaccine Contracts Could Stay Secret," National Public Radio, npr.org, September 29, 2020.
6. Rebecca Robbins, "Moderna, Racing for Profits, Keeps Covid Vaccine Out of Reach of Poor," *NYT*, October 13, 2021.
7. Michael Erman, "Pfizer, Moderna Seen Reaping Billions from COVID-19 Vaccine Booster Market," reuters.com, August 13, 2021.
8. See: "Cuba's COVID Vaccine Rivals BioNTech-Pfizer, Moderna," Deutsche Welle, dw.com, June 27, 2021.
9. "Why Cuba Developed Its Own Covid Vaccine—and What Happened Next," *BMJ*, August 5, 2021.
10. As of October 25, 2021, 100 percent of Cuba's population had received at least one dose of vaccine, and as of January 2022, 86 percent of its population was fully inoculated. The Abdala vaccine, developed by Cuba's public biotech sector, reported 100 percent efficacy in prevention of severe systemic COVID-19 disease and death in a phase 3 trial. Kenny Stancil, "'Historic Turning Point': Cuba Issues Plan for Vaccine Internationalism," Common Dreams, January 25, 2022, https://www.commondreams.org/news/2022/01/25/historic-turning-point-cuba-issues-plan-vaccine-internationalism; Center for Genetic Engineering and Biotechnology, "Abdala Vaccine: 100% Efficacy against Severe Disease and Death in Its Phase III Trial," July 19, 2021, https://www.cigb.edu.cu/en/news/abdala-vaccine-100-efficacy-against-severe-disease-and-death-in-its-phase-iii-trial.
11. See *BMJ*, "Profiteering from Vaccine Inequity: A Crime Against Humanity?" August 16, 2021.

12. "Why Cuba Developed Its Own Covid Vaccine," *BMJ*, August 5, 2021.

13. "Remarks by President Biden on the Drawdown of U.S. Forces in Afghanistan," whitehouse.gov, July 8, 2021.

14. Hina Shamsi, quoted in Asma Khalid, "Biden Pledged to End the Forever Wars, But He Might Just Be Shrinking Them," National Public Radio, September 8, 2021.

15. Quoted in Leo Shane III, "DoD Planned to Spend Billions on Afghan Security Forces. This Group Has a Suggestion for Those Funds," *Military Times*, August 27, 2021.

16. The bill was passed by a bipartisan Senate vote on December 15 and signed into law by Biden on December 27. US Senate Committee on Armed Services, "Reed, Inhofe Praise Senate Passage of National Defense Authorization Act for Fiscal Year 2022," December 15, 2021.

17. See Rich Lowry, "We Need a $1 Trillion Defense Budget," *Ocean Times*, March 4, 2022; and Matthew Kroenig, "Washington Must Prepare for War with Both Russia and China," *Foreign Policy*, February 18, 2022.

18. See Elliott Abrams, "The New Cold War," Council on Foreign Relations, March 4, 2022; and Robert M. Gates, "We Need a More Realistic Strategy for the Post–Cold War Era," *Washington Post*, March 3, 2022.

19. Mike Stone, "Biden Wants $813 Billion for Defense as Ukraine Crisis Raises Alarm," Reuters, March 28, 2022.

20. Davide Castelvecchi, "Ukraine Nuclear Power Plant Attack: Scientists Assess the Risks," *Nature*, March 4, 2022.

21. Chris Hedges, "The Greatest Evil Is War," ScheerPost, February 27, 2022.

22. Emma Newburger, "Biden Administration Proposes Oil and Gas Drilling Reform but Stops Short of Ban," CNBC, November 26, 2021.

23. Cristina Criddle, "Bitcoin Consumes 'More Electricity Than Argentina,'" BBC.com, February 10, 2021.

24. And what about NFTs (non-fungible tokens), you ask? The same absurdity on steroids. See: Luke Savage, "NFTs Are, Quite Simply, Bullshit," *Jacobin*, January 26, 2022.

25. Sohale Andrus Mortazavi, "Cryptocurrency Is a Giant Ponzi Scheme," *Jacobin*, January 21, 2022.

26. Xiaowei Wang, *Blockchain Chicken Farm: And Other Stories of Tech in China's Countryside* (New York: Farrar, Straus, and Giroux, 2020).

27. See Democracy at Work website, "All Things Co-op: Blockchain and Cryptocurrency," February 1, 2022. https://www.democracyatwork.info/atc_blockchain_and_cryptocurrency?utm_campaign=nd_roundup_02022022&utm_medium=email&utm_source=democracyatwork.

Introduction

1. "The Feel-Good Gene," Richard A. Friedman, *NYT*, March 6, 2015.

2. "Global Warming Skeptic Organizations," Union of Concerned Scientists, 2013.

3. See *The Cigarette Papers*, edited by Stanton Glantz et al., 1998.

4. Sheldon Krimsky, *Conflicts of Interest in Science: How Corporate-Funded Academic*

Research Can Threaten Public Health, 2019.

5.   Anahad O'Connor, "Coca-Cola Funds Scientists Who Shift Blame for Obesity Away from Bad Diets," *NYT,* August 9, 2015; Candice Choi, "How Candy Makers Shape Nutrition Science," Associated Press, June 2, 2016.

6.   Farhad Manjoo, "Google, Not the Government, Is Building the Future," *NYT,* May 17, 2017.

7.   Naomi Oreskes and Erik M. Conway, *Merchants of Doubt: How a Handful of Scientists Obscured the Truth on Issues from Tobacco Smoke to Global Warming,* 2010; Sheldon Krimsky, *Conflicts of Interest in Science: How Corporate-Funded Academic Research Can Threaten Public Health,* 2019; Marion Nestle, *Unsavory Truth: How Food Companies Skew the Science of What We Eat,* 2018.

8.   See chapter 9: "Think Tanks and the Betrayal of Reason."

9.   "Donald Trump Flushes Away America's Reputation," *NYT* Editorial Board, January 12, 2018.

10.   Congressional Research Service (Library of Congress), analysis of data from Analytical Perspectives, *Budget of the United States Government, Fiscal Year 2017.*

11.   Evan Ackerman, "A Brief History of the Microwave Oven," *IEEE Spectrum,* September 2016.

12.   Years in the twenty-first century when Sanders voted for war budgets and military appropriations included 2002, 2004, 2005, 2006, 2007, 2008, 2009, 2010, and 2013.

13.   **Iraq:** Iraq Liberation Act, 1998; **Afghanistan:** Authorization for Use of Military Force, 2001; **Kosovo:** Kosovo Resolution, 1999; **Somalia:** Authorization for Use of U.S. Armed Forces in Somalia, 1993.

Chapter 1: The Big Fat Lie

1.   Michele Simon, "Are America's Nutrition Professionals in the Pocket of Big Food?," *Eat Drink Politics,* January 2013. See also the Center for Communicable Disease's website, CDC.gov, for morbidity and mortality statistics regarding heart disease, cancer, stroke, and diabetes.

2.   I owe this chapter's title to Gary Taubes's seminal nutrition science critique "What If It's All Been a Big Fat Lie?," *NYT Magazine,* July 7, 2002.

3.   Anahad O'Connor, "More Evidence That Nutrition Studies Don't Always Add Up," *NYT,* September 29, 2018.

4.   O'Connor, "More Evidence."

5.   National Nutrition Monitoring and Related Research Act of 1990. Public Law 101-445, October 22, 1990.

6.   Dietary Guidelines Advisory Committee, Health.gov, scientific report, 2015.

7.   Victor Oliveira, *The Food Assistance Landscape: FY 2015 Annual Report,* US Department of Agriculture.

8.   Nina Teicholz, "The Scientific Report Guiding the US Dietary Guidelines: Is It Scientific?" *BMJ,* September 23, 2015.

9.   Teicholz, "Scientific Report."

10.   Anahad O'Connor, "How the Sugar Industry Shifted Blame to Fat," *NYT,* September 12, 2016.

11. Cristin E. Kearns, Laura A. Schmidt, and Stanton A. Glantz, "Sugar Industry and Coronary Heart Disease Research: A Historical Analysis of Internal Industry Documents," *JAMA Internal Medicine*, November 2016.

12. A review in a medical journal is a comprehensive, comparative analysis of the influential recent literature on a given topic.

13. Robert B. McGandy, D. M. Hegsted, and F. J. Stare, "Dietary Fats, Carbohydrates and Atherosclerotic Vascular Disease," *New England Journal of Medicine*, July 27, 1967.

14. O'Connor, "How the Sugar Industry Shifted Blame."

15. David Singerman, "The Shady History of Big Sugar," *NYT*, September 16, 2016.

16. Anahad O'Connor, "Coca-Cola Funds Scientists Who Shift Blame for Obesity Away from Bad Diets," *NYT*, August 9, 2015.

17. Email dated August 30, 2014. Associated Press, "Excerpts from Emails between Coke, Anti-obesity Group," November 24, 2015. Emphasis in original.

18. Email dated November 9, 2015. Associated Press, Nov. 24, 2015.

19. David Olinger, "CU Nutrition Expert Accepts $550,000 from Coca-Cola for Obesity Campaign," *Denver Post*, December 26, 2015.

20. O'Connor, "Coca-Cola Funds Scientists."

21. Molly McCluskey, "Public Universities Get an Education in Private Industry," *Atlantic*, April 3, 2017.

22. Quoted by McCluskey, "Public Universities Get an Education."

23. Andrew Jacobs, "How Chummy Are Junk Food Giants and China's Health Officials? They Share Offices," *NYT*, January 9, 2019.

24. Jacobs, "How Chummy Are Junk Food Giants."

25. Andrew Jacobs, "A Shadowy Industry Group Shapes Food Policy around the World," *NYT*, September 16, 2019.

26. Candice Choi, "How Candy Makers Shape Nutrition Science," Associated Press, June 2, 2016.

27. Choi, "How Candy Makers Shape Nutrition Science."

28. Singerman, "Shady History of Big Sugar."

29. Singerman, "Shady History of Big Sugar."

30. Singerman, "Shady History of Big Sugar."

31. Singerman, "Shady History of Big Sugar."

32. Michael Pollan, "Unhappy Meals," *NYT Magazine*, January 28, 2007.

33. *Dietary Goals of the United States*, US Senate Select Committee on Nutrition and Human Needs, 1977.

34. *Diet, Nutrition, and Cancer*, National Academy of Sciences, 1982.

35. Pollan, "Unhappy Meals."

36. Pollan, "Unhappy Meals."

37. Pollan, "Unhappy Meals."

38. Pollan, "Unhappy Meals."

Chapter 2: The Green Revolution

1. "Hunger Statistics," World Food Programme, United Nations. Also see "Maternal and Child Nutrition," *Lancet*, June 6, 2013.

2. Herbert Spencer, *Principles of Biology*, 1864. See also: Darwin Correspondence Project, letter from Darwin to A. R. Wallace, July 5, [1866].

3. Peter Rosset, Joseph Collins, and Frances Moore Lappé, "Lessons from the Green Revolution," *Tikkun*, March/April 2000.

4. For case studies describing the plight of pineapple growers in Ghana and cocoa farmers in Côte d'Ivoire, see: Kojo Amanor, "Global Resource Grabs, Agribusiness Concentration and the Smallholder: Two West African Case Studies," *Journal of Peasant Studies*, July 2012.

5. UNCTAD, "Tracking the Trend towards Market Concentration: The Case of the Agricultural Input Industry," April 20, 2006.

6. Kristina Hubbard, *Out of Hand*, National Family Farm Coalition, December 2009.

7. UNCTAD, "Tracking the Trend towards Market Concentration."

8. Rosset et al., "Lessons from the Green Revolution."

Chapter 3: From Green Revolution to Gene Revolution

1. Mark Derr, "Crossbreeding to Save Species and Create New Ones," *NYT*, July 9, 2002.

2. Sheng Jian Ji et al., "Recombinant Scorpion Insectotoxin AaIT Kills Specifically Insect Cells but Not Human Cells," *Nature*, June 10, 2012.

3. Lucy Sharratt, "Enviropig: A Piggy You Hope Never to Meet at Market," *Common Ground*, October 6, 2011.

4. Quoted in Sharratt, "Enviropig."

5. The percentages are based on USDA statistics of US acreage as of 2015. "The GE Process," Institute for Responsible Technology, 2016.

6. Charles M. Benbrook, "Impacts of Genetically Engineered Crops on Pesticide Use in the U.S.—the First Sixteen Years," *Environmental Sciences Europe*, September 2012.

7. This data, from the United States Geological Survey, was reported in Danny Hakim, "Doubts about the Promised Bounty of Genetically Modified Crops," *NYT*, October 29, 2016.

8. Hakim, "Doubts about the Promised Bounty." For more on the GMO-herbicide relationship, see the section in this chapter "Glyphosate, Monsanto, Roundup, and GMO Safety."

9. Daniel M. Gold, "Review: In 'Food Evolution,' Scientists Strike Back," *NYT*, June 22, 2017. The review is of a pro-GMO documentary film narrated by deGrasse Tyson.

10. European Network of Scientists for Social and Environmental Responsibility, October 21, 2013. See also: Angelika Hilbeck et al., "No Scientific Consensus on GMO Safety," *Environmental Sciences Europe*, January 24, 2015.

11. National Academies of Sciences, Engineering and Medicine, "Report in Brief," *Genetically Engineered Crops: Experiences and Prospects*, May 2016.

12. Sheldon Krimsky and Tim Schwab, "Conflicts of Interest among Committee Members in the National Academies' Genetically Engineered Crop Study," *PLoS ONE*, Febrary 28, 2017.

13. Stephanie Strom, "National Biotechnology Panel Faces Conflict of Interest

Questions," *NYT*, December 27, 2016.

14. Strom, "National Biotechnology Panel."

15. The Big Six as of 2013 were Monsanto, DuPont, Dow, Syngenta, Bayer, and BASF. They have since consolidated into a Big Three or Four.

16. James Ridgeway, "Black Ops, Green Groups," *Mother Jones*, April 11, 2008.

17. Stacy Malkan, "GMO Answers Is a Marketing and PR Website for GMO Companies," US Right To Know, March 26, 2015.

18. Eric Lipton, "Food Industry Enlisted Academics in G.M.O. Lobbying War, Emails Show," *NYT*, September 5, 2015. Emphasis mine.

19. Lipton, "Food Industry Enlisted Academics."

20. Lipton, "Food Industry Enlisted Academics."

21. Lipton, "Food Industry Enlisted Academics."

22. National Academy of Science, "Report in Brief," *Genetically Engineered Crops: Experiences and Prospects*, May 2016.

23. National Academy of Science, "Report in Brief."

24. Hakim, "Doubts about the Promised Bounty."

25. Hakim, "Doubts about the Promised Bounty."

26. Hakim, "Doubts about the Promised Bounty."

27. Becky Price and Janet Cotter, "The GM Contamination Register: A Review of Recorded Contamination Incidents Associated with Genetically Modified Organisms, 1997–2013," *International Journal of Food Contamination*, December 2014. For a representative example of such an incident, see: "Genetically Modified Crop on the Loose and Evolving in U.S. Midwest," *Scientific American*, August 6, 2010.

28. See chapter 7.

29. See, for example: "GMO Corn Is Safe, and Even Has Health Benefits, Analysis of 6,000 Studies Concludes," *Newsweek*, February 22, 2018.

30. "GMO Facts," Non-GMO Project website, 2019. Emphasis added.

31. As the result of a merger in 2018, Monsanto has now been subsumed into Bayer AG.

32. Carey Gillam, *Whitewash: The Story of a Weed Killer, Cancer, and the Corruption of Science* (Washington, DC: Island Press, 2017).

33. Quoted in Carey Gillam, *The Monsanto Papers: Deadly Secrets, Corporate Corruption, and One Man's Search for Justice* (Washington, DC: Island Press, 2021).

34. Monsanto Roundup Lawsuit Update, Ronald V. Miller, Lawsuit Information Center, March 12, 2022. https://www.lawsuit-information-center.com/roundup-mdl-judge-question-10-billion-settlement-proposal.html

35. American Academy of Environmental Medicine, "Genetically Modified Foods," May 8, 2009.

36. A. Fairbrother and R. S. Bennett, "Ecological Risk Assessment and the Precautionary Principle," *Environmental Protection Agency Science Inventory*, 1999.

37. US Environmental Protection Agency, "About Risk Assessment," epa.gov, 2019.

38. Carol Dansereau, *What It Will Take: Rejecting Dead Ends and False Friends in the Fight for the Earth*, 2016.

39. "GMO Facts," Non-GMO Project. Based on "Adoption of Genetically Engi-

neered Crops in the U.S.: Recent Trends in GE Adoption," US Department of Agriculture, Economic Research Service, July 9, 2015.

40. Gary Ruskin, "Seedy Business," US Right to Know, January 2015.

41. "Laureates Letter Supporting Precision Agriculture (GMOs)," June 29, 2016, SupportPrecisionAgriculture.org, "the official site of the Nobel Laureates' pro-GMO campaign."

42. James Dearie, "Plowshares Activists Arrested for Action at Georgia Naval Base," *National Catholic Reporter*, April 5, 2018. See also: Marjorie Cohn, "Threatened with Long Prison Sentences, Anti-nuclear Activists Decry Pentagon 'Brainwashing,'" *Truthout*, October 29, 2019.

43. Jonathan Latham, "107 Nobel Laureate Attack on Greenpeace Traced Back to Biotech PR Operators," *Independent Science News*, July 1, 2016.

44. Boldface emphasis in original.

45. Amy Harmon, "How Square Watermelons Get Their Shape, and Other G.M.O. Misconceptions," *NYT*, August 2, 2016.

46. See the comparative evidence cited in the subsection above headed "But Wait! Do GMO Crops Even Produce More Food?"

Chapter 4: The Tobacco Strategy

1. 1953 magazine advertisement for Chesterfields cigarettes.

2. "Deaths and Mortality Statistics," US Centers for Disease Control and Prevention, 2017; World Health Organization, "Cancer: Key Facts," September 12, 2018.

3. Naomi Oreskes and Erik M. Conway, *Merchants of Doubt*, 2010.

4. *Bad Science: A Resource Book* is available online at the University of California, San Francisco, website.

5. Allan M. Brandt, *The Cigarette Century: The Rise, Fall, and Deadly Persistence of the Product That Defined America*, 2007. Brandt is a professor of the history of medicine at Harvard University.

6. Oreskes and Conway, *Merchants of Doubt*.

7. n.b.: It is a reductionist fallacy to treat complex issues as having only two sides.

8. Hill & Knowlton memo, May 3, 1954. Truth Tobacco Industry Documents, "Report on TIRC Booklet, 'A Scientific Perspective on the Cigarette Controversy.'" See also Michael J. Goodman, "Tobacco's PR Campaign: The Cigarette Papers," *Los Angeles Times*, September 18, 1994.

9. Tens of millions of pages of internal corporate communications are accessible and searchable online at Truth Tobacco Industry Documents.

10. Emphasis in original.

11. "Tobacco-Related Mortality," US Centers for Disease Control and Prevention, 2018.

12. World Health Organization, "Tobacco: Key Facts," May 29, 2019.

13. Oreskes and Conway, *Merchants of Doubt*.

Chapter 5: Fraudulent Pharma

1. Liyan Chen, "The Most Profitable Industries in 2015," *Forbes*, September 23,

2015.

2. Associated Press, August 27, 2019.

3. Scott Higham, Sari Horwitz, and Steven Rich, "76 Billion Opioid Pills: Newly Released Federal Data Unmasks the Epidemic," *Washington Post*, July 16, 2019.

4. Senator Claire McCaskill, quoted in Deirdre Shesgreen, Jayne O'Donnell, and Terry DeMio, "How Drug Company Money Turned Patient Groups into 'Cheerleaders for Opioids'," *USA Today*, February 13, 2018.

5. Jan Hoffman, "Johnson & Johnson Ordered to Pay $572 Million in Landmark Opioid Trial," *NYT*, August 26, 2019.

6. Jan Hoffman, "Purdue Pharma Tentatively Settles Thousands of Opioid Cases," *NYT*, September 11, 2019.

7. Matthew Goldstein, Danny Hakim, and Jan Hoffman, "Sacklers vs. States: Settlement Talks Stumble over Foreign Business," *NYT*, August 30, 2019.

8. The three are AmerisourceBergen, Cardinal Health, and McKesson Corporation, which distribute about 90 percent of the medicines in the United States.

9. Emily Walden, quoted in Spencer Bokat-Lindell, "Making Drug Companies Pay for the Opioid Epidemic," *NYT*, October 22, 2019.

10. Associated Press, "Pfizer Shifting Drug Research to Focus on Diseases with High Profit Potential," October 1, 2008.

11. Associated Press, "Pfizer Shifting Drug Research."

12. "Orphan Products: Hope for People with Rare Diseases," US Food and Drug Administration website, undated.

13. William J. Broad, "Billionaires with Big Ideas Are Privatizing American Science," *NYT*, March 15, 2014.

14. Broad, "Billionaires with Big Ideas."

15. Jake Bernstein, "MIA in the War on Cancer: Where Are the Low-Cost Treatments?" *ProPublica*, April 23, 2014.

16. Michelle D. Holmes et al., "Aspirin Intake and Survival after Breast Cancer," *Journal of Clinical Oncology*, February 16, 2010.

17. Quoted in Bernstein, "MIA in the War on Cancer."

18. A standard euphemism for the world's economically deprived populations is "developing countries," an ideologically loaded term reflecting the neoliberal myth that free markets are actually solving the problem of economic inequality. "Underdeveloped countries" is more honest, but awkward. I prefer "Third World," which is technically outdated due to the demise of the "Second World" (the Soviet bloc), but is still a serviceable and more neutral term.

19. Rick Gladstone, "W.H.O. Assails Delay in Ebola Vaccine," *NYT*, November 3, 2014.

20. Quoted in Naki B. Mendoza, "A New Look for Big Pharma Business in the Developing World," Devex.com, April 7, 2016.

21. "Big Win for Affordable Medicine," *BloombergQuint*, April 30, 2019.

22. Joseph E. Stiglitz, "Don't Trade Away Our Health," *NYT*, January 31, 2015.

23. R. Jai Krishna and Jeanne Whalen, "Novartis Loses Glivec Patent Battle in India," *Wall Street Journal*, April 1, 2013.

24. Simon Reid-Henry and Hans Lofgren, "Pharmaceutical Companies Putting Health of World's Poor at Risk," *Guardian*, July 26, 2012.

25. Jack Ellis, "Novartis Is the Latest Big Pharma Player to Extend Olive Branch in Third World Patent Rights Row," Intellectual Asset Management, September 15, 2015.

26. Michael Edwards, "R&D in Emerging Markets: A New Approach for a New Era," McKinsey and Company, February 2010. McKinsey and Company was in the headlines in July 2019, when a lawsuit revealed it had advised a pharmaceutical company to "get more patients on higher doses of opioids" and to devise ways "for keeping patients on opioids longer." Walt Bogdanich, "McKinsey Advised Johnson & Johnson on Increasing Opioid Sales," *NYT*, July 25, 2019.

27. Gardiner Harris, "Maker of Costly Hepatitis C Drug Sovaldi Strikes Deal on Generics for Poor Countries," *NYT*, September 15, 2014.

28. Julie Schmit, "Costs, Regulations, Move More Drug Tests Outside USA," *USA Today*, May 16, 2005.

29. Joe Stephens, "Panel Faults Pfizer in '96 Clinical Trial in Nigeria," *Washington Post*, May 7, 2006.

30. As of December 31, 2017 (Gates Foundation Factsheet).

31. Associated Press, "With Poo on a Pedestal, Bill Gates Talks Toilets," November 6, 2018.

32. Michelle Goldberg, "Melinda Gates' New Crusade: Investing Billions in Women's Health," *Newsweek*, May 7, 2012.

33. Paul B. Farrell, "Gates's $4 Billion Foray in Global Family Planning," *MarketWatch*, May 15, 2012.

34. Charles Ornstein and Katie Thomas, "Top Cancer Researcher Fails to Disclose Corporate Financial Ties in Major Research Journals," *NYT*, September 8, 2018.

35. "Medicine's Financial Contamination," *NYT* Editorial Board, September 14, 2018.

36. Marcia Angell, "Drug Companies and Doctors: A Story of Corruption," *New York Review of Books*, January 15, 2009.

37. Angell, "Drug Companies and Doctors."

38. Evidence of the ubiquity of conflicts of interest in medical science has been patiently assembled over many years by medical ethicist Sheldon Krimsky. See especially his books *Science in the Private Interest* (2003) and *Conflicts of Interest in Science* (2019).

39. Angell, "Drug Companies and Doctors."

40. Personal correspondence (marked "Confidential"), Charles B. Nemeroff to Thomas J. Lawley, May 22, 2000. This letter can be accessed on the US Senate Committee on Finance website. SmithKline Beecham merged with Glaxo Wellcome in 2000 to form GlaxoSmithKline.

41. Gardiner Harris, "Top Psychiatrist Didn't Report Drug Makers' Pay," *NYT*, October 3, 2008.

42. Katie Thomas and Michael S. Schmidt, "Glaxo Agrees to Pay $3 Billion in Fraud Settlement," *NYT*, July 2, 2012.

43. Martha Rosenberg, "Big Pharma Gone Wild," AlterNet, February 3, 2009. See also Dyan Neary, "We Need to Talk about Frankie," *The Cut*, April 24, 2017. Risperdal was initially approved by the FDA to be marketed for use by adults but not by children. It has been nonetheless prescribed extensively for children with

psychological and/or behavioral issues, because once a drug is on the market it is not illegal to prescribe it "off-label." Risperdal now has several FDA-approved pediatric indications in spite of its known track record of causing permanent tardive dyskinesia (an involuntary movement disorder), excessive weight gain, the growth of breasts in boys and men, severe drooling, and respiratory issues.

44. Gardiner Harris, "Drug Maker Told Studies Would Aid It, Papers Say," *NYT*, March 20, 2009.

45. Llewellyn Hinkes-Jones, "Bad Science," *Jacobin*, June 1, 2014.

46. Katie Thomas, "Furor Over Drug Prices Puts Patient Advocacy Groups in Bind," *NYT*, September 27, 2016.

47. Andreas Zumach, "Who Is Really Helping the WHO?", *DW* (Deutsche Welle), May 21, 2012.

48. Carl Elliott, "Useless Studies, Real Harm," *NYT*, July 28, 2011.

49. Elliott, "Useless Studies, Real Harm."

50. Elliott, "Useless Studies, Real Harm."

51. Quoted in Sarah Boseley, "Scandal of Scientists Who Take Money for Papers Ghostwritten by Drug Companies," *Guardian*, February 7, 2002.

52. Melody Petersen, "Madison Ave. Has Growing Role in the Business of Drug Research," *NYT*, November 22, 2002.

53. Dr. Arnold S. Relman, professor emeritus at Harvard Medical School. Quoted in Petersen, "Madison Ave. Has Growing Role."

54. Kent Sepkowitz, "Under the Influence? Drug Companies, Medical Journals, and Money," *Slate*, September 10, 2007.

55. Richard Horton, "The Dawn of McScience," *New York Review of Books*, March 11, 2004.

56. A. J. Wakefield et al., "Ileal-lymphoid-nodular Hyperplasia, Non-specific Colitis, and Pervasive Developmental Disorder in Children," *Lancet*, 1998.

57. See: L. E. Taylor et al., "Vaccines Are Not Associated with Autism: An Evidence-Based Meta-Analysis of Case-Control and Cohort Studies," *Vaccine*, June 17, 2014; and Anders Hviid et al., "Measles, Mumps, Rubella Vaccination and Autism: A Nationwide Cohort Study," *Annals of Internal Medicine*, April 16, 2019.

58. T. S. Sathyanarayana and Chittaranjan Andrade, "The MMR Vaccine and Autism: Sensation, Refutation, Retraction, and Fraud," *Indian Journal of Psychiatry*, April–June 2011.

59. Rebecca Robbins, "Meeting with Trump Emboldens Antivaccine Activists, Who See an Ally in the Oval Office," Statnews.com, November 30, 2016.

60. "Vaccine Hesitancy: A Generation at Risk," *Lancet Child & Adolescent Health*, May 2019.

61. Centers for Disease Control and Prevention, "CDC Media Statement: Measles Cases in the U.S. Are Highest since Measles Was Eliminated in 2000," April 25, 2019.

62. Cenyu Shen and Bo-Christer Björk, "'Predatory' Open Access: A Longitudinal Study of Article Volumes and Market Characteristics," *BMC Medicine*, October 1, 2015.

63. Gina Kolata, "How to Report When the Science Is Sketchy," *NYT*, October 30,

2017.

64. Gina Kolata, "Many Academics Are Eager to Publish in Worthless Journals," *NYT*, October 31, 2017.

65. Angell, "Drug Companies and Doctors."

66. Lynn Payer, *Disease-Mongers: How Doctors, Drug Companies, and Insurers Are Making You Feel Sick*, 1992.

67. Selective serotonin reuptake inhibitors.

68. See: Paula J. Caplan, "Premenstrual Mental Illness: The Truth about Serafem," *Network News*, National Women's Health Network, 2001.

69. See: Brendan I. Koerner, "Disorders Made to Order," *Mother Jones*, 2002.

70. Quoted in Angell, "Drug Companies and Doctors."

71. See: Charles Barber, "Are We Really So Miserable?," *Salon*, August 26, 2009. Barber points out that aside from the United States, the only country in the world where direct-to-consumer drug advertising is not illegal is New Zealand.

72. Paula J. Caplan, "Pathologizing Your Period," *Ms.*, 2008.

73. Natasha Singer and Duff Wilson, "Menopause, as Brought to You by Big Pharma," *NYT*, December 13, 2009.

74. Bruce E. Levine, "Moody Is the New Bipolar," *AlterNet*, November 14, 2007.

75. Steven Woloshin and Lisa M. Schwartz, "Giving Legs to Restless Legs: A Case Study of How the Media Helps Make People Sick," *PLoS Medicine*, April 11, 2006.

76. Koerner, "Disorders Made to Order."

77. Koerner, "Disorders Made to Order."

78. Sheldon Krimsky, *Science in the Private Interest*, 2003.

79. Dennis Cauchon, "FDA Advisers Tied to Industry," *USA Today*, September 25, 2000.

80. Marc Kaufman, "Many Workers Call FDA Inadequate at Monitoring Drugs," *Washington Post*, December 17, 2004. The federal agency did not intend to make the results of its investigation public; the report was forced into the open through the efforts of two science information groups: Public Employees for Environmental Responsibility and the Union of Concerned Scientists.

81. Russell Mokhiber and Robert Weissman, "Dr. Gottlieb Is Not Happy," *Common Dreams*, August 2, 2006. Dr. Scott Gottlieb was the Trump administration's FDA commissioner from May 2017 through April 2019.

82. Mokhiber and Weissman, "Dr. Gottlieb Is Not Happy."

83. "Legalized Bribery Alive and Well in Washington as Congress Reapproves FDA User Fees for Drug Companies," *Off the Grid News*, 2012.

84. Gillian Woollett and Jay Jackson, "FDA Has Received $7.67 Billion from Manufacturers to Fund Drug Review," Avalere Health press release, August 15, 2016.

Chapter 6: Spitting in the Well We Drink From

1. Mona Hanna-Attisha, "I Helped Expose the Lead Crisis in Flint. Here's What Other Cities Should Do," *NYT*, August 28, 2019.

2. Joseph M. Leonard and William J. Gruhn, "Flint River Assessment," Michigan Department of Natural Resources, July 2001.

3. Emma Winowiecki, "Does Flint Have Clean Water? Yes, but It's Complicated,"

Michigan Radio (NPR), April 25, 2019.

4. The prosecutors dropped those charges in June 2019, but insisted they did so only to clear the way for new charges based on stronger evidence.

5. Colby Itkowitz, "The Heroic Professor Who Helped Uncover the Flint Lead Water Crisis, Has Been Asked to Fix It," *Washington Post*, January 27, 2016.

6. Itkowitz, "Heroic Professor."

7. Quoted in Katie L. Burke, "Flint Water Crisis Yields Hard Lessons in Science and Ethics," *American Scientist*, May–June 2016.

8. Nick Corasaniti, "Newark Water Crisis: Racing to Replace Lead Pipes in Under Two Years," *NYT*, August 26, 2019.

9. Hanna-Attisha, "I Helped Expose the Lead Crisis in Flint."

10. David Pimentel, "Ecology of Increasing Diseases: Population Growth and Environmental Degradation," *Human Ecology*, October 2007.

11. See: Michael Egan, *Barry Commoner and the Science of Survival: The Remaking of Environmentalism*, 2007.

12. John M. Lee, "Silent Spring Is Now Noisy Summer: Pesticides Industry Up in Arms," *NYT*, July 22, 1962.

13. Carol Dansereau, *What It Will Take: Rejecting Dead Ends and False Friends in the Fight for the Earth*, 2016.

14. The quotations in this paragraph are from Nixon's 1970 State of the Union address.

15. "US Science Agencies Face Deep Cuts in Trump Budget," *Nature*, March 16, 2017.

16. Timothy M. O'Donnell, "Of Loaded Dice and Heated Arguments," *Social Epistemology*, 2000.

17. "CO_2 Concentrations Hit Highest Levels in 3 Million Years," *E360 Digest*, Yale University, May 14, 2019.

18. "Extreme Weather: National Climate Assessment," US Global Change Research Program, May 2014.

19. Of the many books that have been written about Katrina, I recommend especially Gary Rivlin's retrospective *Katrina: After the Flood* (2016) and Naomi Klein's *The Shock Doctrine: The Rise of Disaster Capitalism* (2008).

20. Brian Kahn, "Looking for Global Warming? Check the Ocean," Climate Central, April 21, 2015.

21. Center for Biological Diversity, "Global Warming and Life on Earth," February 27, 2017.

22. For one of innumerable authoritative sources, see: Intergovernmental Panel on Climate Change, *IPCC Fifth Assessment Report*, 2013.

23. Suzanne Goldenberg, "Revealed: The Day Obama Chose a Strategy of Silence on Climate Change," *Guardian*, November 1, 2012.

24. Dana Nuccitelli, "Trump Is Copying the Bush Censorship Playbook. Scientists Aren't Standing for It," *Guardian*, January 31, 2017.

25. This and the following quotations in this section are quoted in Dansereau, *What It Will Take*. The speeches they are from are available in video and transcript form at Shallownation.com. See "President Obama New Mexico Speech Video: Mar. 21, 2012, Oil Fields near Maljamar" and "President Obama Cushing, OK

Speech Video: Mar. 22, 2012, Keystone XL Pipeline Announcement."

26. See the previous note.

27. James Lawrence Powell, "The Consensus on Anthropogenic Global Warming," *Skeptical Inquirer*, November/December 2015. Powell served for twelve years on the National Science Board in the Reagan and GHW Bush administrations. The peer-reviewed articles he investigated were published in 2013 and 2014.

28. Donald Trump, Twitter, @*realDonaldTrump*, January 1, 2014.

29. James E. Hansen, in an interview with Canadian Broadcasting Corporation News, April 27, 2013.

30. Coral Davenport, "Climate Change Denialists in Charge," *NYT*, March 27, 2017.

31. "Pruitt v. EPA: 14 Challenges of EPA Rules by Oklahoma Attorney General," *NYT*, January 14, 2017.

32. Jeff Desjardins, "The Oil Market Is Bigger Than All Metal Markets Combined," *Visual Capitalist*, October 14, 2016.

33. "Global Coal Industry 2013–2018: Trends, Profits and Forecast Analysis," *Research and Markets*, April 2014.

34. Neela Banerjee, Lisa Song, and David Hasemyer, "Exxon's Own Research Confirmed Fossil Fuels' Role in Global Warming Decades Ago," *Inside Climate News*, September 16, 2015.

35. Banerjee et al., "Exxon's Own Research Confirmed."

36. Banerjee et al., "Exxon's Own Research Confirmed."

37. Quoted in Justin Gillis and Clifford Krauss, "Exxon Mobil Investigated for Possible Climate Change Lies by New York Attorney General," *NYT*, November 5, 2015.

38. Documented by Greenpeace, using annual IRS filings by the Koch family foundations. The Koch brothers had been generally successful in hiding their right-wing funding operations from public view until 2010, when investigative journalist Jane Mayer exposed it in "Covert Operations: The Billionaire Brothers Who Are Waging a War against Obama," *New Yorker*, August 30, 2010. David Koch died in August 2019.

39. Forbes/Profile/Koch Family, Forbes.com, October 6, 2016.

40. "Global Warming Skeptic Organizations," Union of Concerned Scientists, August 2013.

41. See: "Trouble in the Heartland," *Economist*, February 15, 2012.

42. Quoted in Tim Dickinson, "Inside the Koch Brothers' Toxic Empire," *Rolling Stone*, September 24, 2014.

43. "Randianism As It Is: Ayn Rand, the Koch Brothers and the Libertarian Party," Counter Currents, July 27, 2015.

44. Joel Kovel, "The Future Will Be Ecosocialist, Because Without Ecosocialism There Will Be No Future," in *Imagine: Living in a Socialist USA*, 2014.

45. Three particularly valuable expositions of the ecosocialist point of view are Carol Dansereau, *What It Will Take* (2016); Chris Williams, *Ecology and Socialism* (2010); and Ian Angus, *A Redder Shade of Green* (2017).

46. Naomi Klein, *This Changes Everything: Capitalism vs. the Climate*, 2014.

47. Dickinson, "Inside the Koch Brothers' Toxic Empire."

48. Nicholas Confessore, "Koch Brothers' Budget of $889 Million for 2016 Is on Par with Both Parties' Spending," *NYT*, January 26, 2015.

49. Dansereau, *What It Will Take*.

50. Sheldon Krimsky, "The Unsteady State and Inertia of Chemical Regulation Under the US Toxic Substances Control Act," *PLoS Biology*, December 2017.

51. Government Accountability Office, "Report to the Congress: High-Risk Series: An Update," January 2009. A subsequent GAO report showed that the EPA's lack of adequate scientific information continued: Government Accountability Office, "Toxic Substances: Report to Congressional Requesters," March 2013.

52. Lisa Heinzerling, quoted by Rachel Aviv, "A Valuable Reputation," *New Yorker*, February 10, 2014.

53. Quoted by Dansereau, *What It Will Take*.

54. Steve Eder, "Neomi Reo, the Scholar Who Will Help Lead Trump's Regulatory Overhaul," *NYT*, July 9, 2017. Reo, a right-wing legal scholar, was named by Trump to head OIRA.

55. Gary Kroll, "Rachel Carson's *Silent Spring:* A Brief History of Ecology as a Subversive Subject," Online Ethics Center, 2002.

56. Tom Gjelten, "Who's Looking at Natural Gas Now? Big Oil," National Public Radio, September 23, 2009.

57. Bryan Walsh, "Exclusive: How the Sierra Club Took Millions from the Natural Gas Industry—and Why They Stopped," *Time*, February 2, 2012.

58. John Schwartz and Brad Plumer, "The Natural Gas Industry Has a Leak Problem," *NYT*, June 21, 2018.

59. Ramón A. Alvarez et al., "Assessment of Methane Emissions from the U.S. Oil and Gas Supply Chain," *Science*, July 13, 2018.

60. Schwartz and Plumer, "The Natural Gas Industry Has a Leak Problem."

61. Hiroko Tabuchi, "Despite Their Promises, Giant Energy Companies Burn Away Vast Amounts of Natural Gas," *NYT*, October 17, 2019.

62. Tabuchi, "Despite Their Promises."

63. Lisa Friedman and Coral Davenport, "Curbs on Methane, Potent Greenhouse Gas, to Be Relaxed in U.S.," *NYT*, August 29, 2019.

64. US Energy Information Administration, "Hydraulically Fractured Wells Provide Two-Thirds of U.S. Natural Gas Production," May 5, 2016.

65. US Energy Information Administration, "Hydraulic Fracturing Accounts for About Half of Current U.S. Crude Oil Production," March 15, 2015.

66. US Geological Survey, "USGS FAQs," November 16, 2016.

67. US Energy Information Administration, "Hydraulically Fractured Wells Provide Two-Thirds of U.S. Natural Gas Production."

68. US Geological Survey, "Induced Earthquakes Raise Chances of Damaging Shaking in 2016," March 28, 2016.

69. Stanley Reed, "Earthquakes Are Jolting the Netherlands. Gas Drilling Is to Blame," *NYT*, October 24, 2019.

70. The long-term average from 1978 to 2008 was two such earthquakes per year in Oklahoma. US Geological Survey, "Record Number of Oklahoma Tremors Raises Possibility of Damaging Earthquakes," May 6, 2014.

71. A terawatt is a trillion watts. David L. Chandler, "Shining Brightly," *MIT News*, October 26, 2011; V. Prema and K Urna Rao, "Predictive Models for Power Management of a Hybrid Microgrid," International Conference on Advances in Energy Conversion Technologies, *IEEE*, 2014.

72. US Department of Energy, "Basic Research Needs for Solar Energy Utilization," DOE Office of Science, April 2005.

73. US Department of Energy, "Basic Research Needs for Solar Energy Utilization."

74. US Energy Information Administration, "U.S. Electricity Generation by Energy Source," EIA.gov, October 25, 2019.

75. Paul Krugman, "Wind, Sun and Fire," *NYT*, February 1, 2016; and "Earth, Wind and Liars," *NYT*, April 16, 2018.

76. Krugman, "Wind, Sun and Fire."

77. Tom Randall, "Wind and Solar Are Crushing Fossil Fuels," *Bloomberg*, April 16, 2016.

78. Shakuntala Makhijani, "Cashing in on All of the Above: U.S. Fossil Fuel Production Subsidies Under Obama," Oilchange International, July 2014.

79. *Reforming Subsidies Could Help Pay for a Clean Energy Revolution*, International Institute for Sustainable Development, August 8, 2019.

80. National Academies of Sciences, Engineering, and Medicine, *Valuing Climate Damages: Updating Estimation of the Social Cost of Carbon Dioxide*, National Academies Press, 2017.

81. Amir Jina, Energy Policy Institute at the University of Chicago, "The $200 Billion Fossil Fuel Subsidy You've Never Heard Of," *Forbes/Energy*, February 1, 2017.

82. Gallup Poll, "U.S. Concern About Global Warming at Eight-Year High," March 16, 2016.

Chapter 7: Atoms for Peace?

1. Dwight D. Eisenhower, "Atoms for Peace," December 8, 1953.

2. Peter Kuznick, "Japan's Nuclear History in Perspective: Eisenhower and Atoms for War and Peace," *Bulletin of the Atomic Scientists*, April 13, 2011.

3. Kuznick, "Japan's Nuclear History."

4. Michael Barletta, "Pernicious Ideas in World Politics: 'Peaceful Nuclear Explosives,'" Monterey Institute of International Studies, September 2000.

5. Quoted in Richard G. Hewlitt and Jack M. Holl, *Atoms for Peace and War: Eisenhower and the Atomic Energy Commission*, 1989. Emphasis added.

6. Quoted in Barletta, "Pernicious Ideas in World Politics."

7. Edward Teller, Wilson K. Talley, Gary H. Higgins, and Gerald W. Johnson, *The Constructive Use of Nuclear Explosions*, 1968.

8. James Hansen et al., "To Those Influencing Environmental Policy but Opposed to Nuclear Power," *NYT*, November 3, 2013.

9. Mark Fischetti, "How to Power the World without Fossil Fuels," *Scientific American*, April 15, 2013.

10. Laura Beans, "300+ Groups Urge Climate Scientist Dr. Hansen to Rethink Support of Nuclear Power," EcoWatch, January 9, 2014.

11. "Chernobyl: The True Scale of the Accident," International Atomic Energy

Agency, September 5, 2015; Elisabeth Cardis et al., "Estimates of the Cancer Burden in Europe from Radioactive Fallout from the Chernobyl Accident," *International Journal of Cancer*, April 20, 2006. Although these figures are not entirely commensurable (because not all cancers result in death), they nonetheless represent sharply conflicting appraisals of the harm caused by the radiation.

12. Alexander R. Sich, "Truth Was an Early Casualty," *Bulletin of the Atomic Scientists*, May 1996.
13. The explosion was not of the atomic bomb type. The meltdown of the overheated reactor created extreme steam pressure that caused the plant to blow up.
14. Steven Starr, "Costs and Consequences of the Fukushima Daiichi Disaster," Environmental Health Policy Institute, Physicians for Social Responsibility, October 31, 2012.
15. *The Fukushima Daiichi Accident*, International Atomic Energy Agency, 2015.
16. Fred Pearce, "What Was the Fallout from Fukushima?" *Guardian*, June 3, 2018. See also: Fred Pearce, *Fallout: Disasters, Lies and the Legacy of the Nuclear Age*, 2018.
17. Linda Sieg, "Japan Ponders Fukushima Options, but Tepco Too Big to Fail," Reuters, September 1, 2013.
18. Helen Caldicott, *Nuclear Madness: What You Can Do*, 1994.
19. *Congressional Quarterly*, "Comprehensive Nuclear Waste Plan Enacted," 1982.
20. "Strategy for the Management and Disposal of Used Nuclear Fuel and High-Level Radioactive Waste," US Department of Energy, January 2013.
21. "Radioactive Wastes—Myths and Realities," and "Safety of Nuclear Power Reactors," World Nuclear Association, February 2016.
22. "Coming Clean About Nuclear Power," *Scientific American*, June 1, 2011.
23. According to NationMaster, in 2003, France's CO_2 emissions (in thousands of metric tonnes per 1,000 people) amounted to 5.84, compared to the US's 19.86. Taking the countries' populations into account, France's total CO_2 emissions in metric tonnes was 362,080,000 compared to the US's 3,773,400,000.
24. Quoted in "Nuclear Energy Risk Management," House Science, Space, and Technology Committee hearing, May 3, 2011.
25. Doug Koplow, "Nuclear Power: Still Not Viable without Subsidies," Union of Concerned Scientists, February 2011.
26. US Nuclear Regulatory Commission, "Backgrounder on Nuclear Insurance and Disaster Relief," June 2014.
27. David Hochschild, quoted in Giles Parkinson, "The Myth about Renewable Energy Subsidies," *Clean Technica*, February 25, 2016.
28. Matthew L. Wald, "Energy Dept. Is Told to Stop Collecting Fee for Nuclear Waste Disposal," *NYT*, November 19, 2013. Emphasis added.
29. "Disposal of High-Level Nuclear Waste," US Government Accountability Office (undated).

Chapter 8: The Academic-Industrial Complex

1. Barbara J. Culliton, "The Academic-Industrial Complex," *Science*, May 28, 1982.
2. Culliton, "Academic-Industrial Complex."

3. See the section in chapter 5 entitled "A Turning Point: The Bayh-Dole Act."
4. Culliton, "Academic-Industrial Complex."
5. Culliton, "Academic-Industrial Complex."
6. Culliton, "Academic-Industrial Complex." Emphasis added.
7. Culliton, "Academic-Industrial Complex."
8. Jeffrey Mervis, "Data Check: U.S. Government Share of Basic Research Funding Falls Below 50%," *Science*, March 9, 2017.
9. Molly McCluskey, "Public Universities Get an Education in Private Industry," *Atlantic*, April 3, 2017.
10. Eileen Buss, quoted by McCluskey, "Public Universities Get an Education."
11. Graham Bowley, "The Academic-Industrial Complex," *NYT*, August 2, 2010.
12. Bowley, "Academic-Industrial Complex."
13. Rensselaer Polytechnical Institute, "Rensselaer Part of Nationwide Effort to Advance Manufacturing Biopharmaceuticals," 2017.
14. Shirley Ann Jackson, "Op-Ed: The New Polytechnic: Preparing to Lead in the Digital Economy," *U.S. News and World Report*, September 22, 2014.
15. Rensselaer Polytechnical Institute, "Rensselaer Part of Nationwide Effort."
16. Meredith Hoffman, "America's Highest Paid College President Is Dragging Her School into Crippling Debt," *Vice News*, December 16, 2017.
17. David L. Kirp, quoted by Felicia R. Lee, "Academic Industrial Complex," *NYT*, September 6, 2003.
18. Dr. Drew Faust resigned as Harvard president on June 30, 2018.
19. Tracy Jan, "Grants for Research Get Scarcer," *Boston Globe*, September 23, 2014.
20. Quoted in Jan, "Grants for Research Get Scarcer."
21. Joaquin Palomino, "Billions of Corporate Dollars Are Hijacking University Research to Help Make Profits," alternet.org, April 22, 2013.
22. Palomino, "Billions of Corporate Dollars."
23. Palomino, "Billions of Corporate Dollars."
24. Palomino, "Billions of Corporate Dollars."
25. Ignacio Chapela, quoted by Palomino.
26. Robert Birgeneau, quoted by Palomino.
27. Mark Yudof, quoted by Palomino.
28. Jason Del Gandio, "Neoliberalism and the Academic Industrial Complex," *Truthout*, August 12, 2010.

Chapter 9: Think Tanks and the Betrayal of Reason

1. Eric Lipton and Brooke Williams, "Scholarship or Business? Think Tanks Blur the Line," *NYT*, August 8, 2016.
2. Jason Stahl, *Right Moves: The Conservative Think Tank in American Political Culture Since 1945*, 2016. Emphasis added.
3. The RAND Corporation is the subject of chapter 21.
4. Quoted by Tevi Troy, "Devaluing the Think Tank," *National Affairs*, Winter 2012.
5. Troy, "Devaluing the Think Tank."
6. "The Mont Pelerin Society: The Ultimate Neoliberal Trojan Horse," *Daily Knell*, October 29, 2012.

7. The influence of these economists will be discussed in chapter 10.
8. For more on the Kochs, see chapters 6 and 10.
9. Ade Adenji, "Koch Money on Campus: Who's Getting Grants and For What?" *Inside Philanthropy*, November 3, 2014.
10. "Donor Influence at George Mason Finally Exposed," unkochmycampus.org, May 2018.
11. Matthew Barakat, "George Mason University Becomes a Favorite of Charles Koch," Associated Press, April 1, 2016.
12. Marshall Steinbaum, "The Book That Explains Charlottesville," *Boston Review*, August 14, 2017.
13. "Toxic Shock," *Economist*, May 26, 2012.
14. See Robert Littlemore, "Heartland Insider Exposes Institute's Budget and Strategy," desmogblog.com, February 14, 2012.
15. Heartland Institute, "Confidential Memo: 2012 Heartland Climate Strategy," January 2012 (attachment to Littlemore, "Heartland Insider Exposes Institute's Budget and Strategy").
16. "The Heartland Institute 2012 Fundraising Plan," January 15, 2012.
17. See the Think Tanks and Civil Societies Program's home page at gotothinktank.com.
18. James G. McGann, "2016 Global Go To Think Tank Index Report," Think Tanks and Civil Society Program, University of Pennsylvania, January 26, 2017.
19. Lipton and Williams, "Scholarship or Business?"
20. Eric Lipton, Nicholas Confessore, and Brooke Williams, "Top Scholars or Lobbyists? Often It's Both," *NYT*, August 9, 2016.

Chapter 10: The Dismal Science Is Certainly Dismal, but Is It Science?

1. Sandra Harding, *The Science Question in Feminism*, 1986.
2. Keynes first used this phrase well before the Great Depression, but he continued to utilize it to deflect all questions about "the long run." John Maynard Keynes, *A Tract on Monetary Reform*, 1923.
3. Hayek is also identified with the Austrian School of economic thought, which is sometimes characterized as a rival to the Chicago School, but these distinctions are beyond the scope of this book.
4. The civil rights movement was an equally important accelerant.
5. Irving Kristol, *Neoconservatism: The Autobiography of an Idea*, 1995.
6. A leading proponent of James M. Buchanan's Public Choice theory declared, "There are no human rights . . . when property rights are not used as the standard." Murray Rothbard, *The Ethics of Liberty*, 1998.
7. Harvard political scientist Theda Skocpol is a leading expert on the Koch network. As an entry point into her work on the subject, see: Theda Skocpol and Alexander Hertel-Fernandez, "The Koch Network and Republican Party Extremism," *Perspectives on Politics*, September 2016.
8. Kevin Robillard, "Koch Network Ramps Up Political Spending While Trying to Push Trump Team," *Politico*, June 24, 2017.

9. Robillard, "Koch Network Ramps Up Political Spending."

10. Katie Glueck, "Trump's Shadow Transition Team," *Politico*, November 22, 2016.

11. Grover Norquist, in an interview on National Public Radio's *Morning Edition*, May 25, 2001.

12. Quoted in Nancy MacLean, *Democracy in Chains: The Deep History of the Radical Right's Stealth Plan for America*, 2017.

13. James M. Buchanan, *The Limits of Liberty: Between Anarchy and Leviathan*, 2000. Quoted in MacLean, *Democracy in Chains*.

14. James McGill Buchanan, *Property as a Guarantor of Liberty*, 1993.

15. MacLean, in *Democracy in Chains*, is but one author to make that comparison. The notion of Buchanan as a "Lenin of libertarianism" originated in a private missive by Murray Rothbard that borrowed its title from Lenin's *What Is To Be Done?* See "Rothbard's Confidential Memorandum to the Volker Fund," July 1961.

16. James M. Buchanan, "Working Papers for Internal Discussion Only," December 1956. Quoted in MacLean, *Democracy in Chains*. In those papers, Buchanan stated that the center was named after Thomas Jefferson to deflect attention from the "extreme views" that were "the real purpose of the program."

17. James M. Buchanan, "America's Third Century," *Atlantic Economic Journal*, November 1973. Quoted in MacLean, *Democracy in Chains*.

18. *Wall Street Journal*, March 19, 1988. Cited by MacLean, *Democracy in Chains*. The Koch Foundation also donated $10 million to GMU's law school in 2016, which subsequently renamed itself the Antonin Scalia School of Law. To the university's dismay, it quickly became known by the acronym ASSLaw. An attempt to change the name to Scalia Law School had little effect: it seems destined to be forever known as ASSLaw on the internet.

19. For more on mathematical game theory, see chapter 21.

20. MacLean, *Democracy in Chains*.

Chapter 11: Science Harnessed to the Chariot of Destruction

1. The title of this chapter is derived from an epigraph in Annie Jacobsen's *Operation Paperclip*, a study of Nazi scientists during and after World War II: "The scale on which science and engineering have been harnessed to the chariot of destruction in Germany is indeed amazing."

2. American Association for the Advancement of Science (AAAS) R&D report series, based on US Office of Management and Budget and agency R&D budget data. Amounts in constant 2016 dollars.

3. Dan Steinbock, "The Challenges for America's Defense Innovation," Information Technology and Innovation Foundation (ITIF), November 20, 2014. n.b.: "Defense" is a propagandistic adjective that I purposefully avoid, using "military budget" or "war industry" instead. I will, however, denote the US Department of Defense by its official name and its acronym, DoD.

4. Scott Wong and Rebecca Kheel, "Mattis: 'I Need to Make the Military More Lethal,'" TheHill.com, February 1, 2018.

5. William D. Hartung, "The Pentagon Budget Still Rising, 40 Years Later," Tom-

Dispatch, December 17, 2019.

6.   Associated Press, "Deal Sealed on Federal Budget Ensures No Shutdown, Default," *NYT,* July 22, 2019.

7.   A careful analysis by William D. Hartung and Mandy Smithberger puts the total at $1.2542 trillion. See Hartung and Smithberger, "Boondoggle, Inc.: Making Sense of the $1.25 Trillion National Security State Budget," TomDispatch, May 7, 2019.

8.   See the section in chapter 20 headed "Trump's 'Space Force' Initiative."

9.   See chapter 7.

10. *Sixty Minutes,* May 12, 1996.

11. Chalmers Johnson, "Empire of Bases," *NYT,* July 13, 2009.

12. Inspector General, US Department of Defense, "Army General Fund Adjustments Not Adequately Documented or Supported," dodig.mil, July 26, 2016.

13. Inspector General, DoD, "Army General Fund Adjustments."

14. Nick Turse, "How Many Wars Is the US Really Fighting?" *Nation,* September 24, 2015. See also: Jeremy Scahill, *Dirty Wars,* 2013.

15. See chapter 18.

16. William Kristol and Robert Kagan, "Toward a Neo-Reaganite Foreign Policy," *Foreign Affairs,* July/August 1996.

17. George W. Bush had not yet eclipsed his brother's political star.

18. Project for the New American Century [PNAC], "Statement of Principles," June 3, 1997.

19. PNAC, "Statement of Principles."

20. The United States Commission on National Security/21st Century, *New World Coming,* September 15, 1999.

21. Quoted in Madeleine Albright's memoir, *Madam Secretary,* 2003.

22. The PPI is an affiliate of the Democratic Leadership Council, the dominant force within the Democratic Party since the 1990s.

23. PNAC, "Open Letter to President Bill Clinton," January 26, 1998.

24. Kimberly Amadeo, "War on Terror Facts, Costs and Timeline," TheBalance.com, October 9, 2017.

25. "The Pentagon Is Not a Sacred Cow," *NYT* Editorial Board, December 13, 2017.

26. Cheryl Gay Stolberg, "Senate Passes $700 Billion Pentagon Bill, More Money Than Trump Sought," *NYT,* September 18, 2017.

27. John M. Donnelly, "Defense to Get Historically High Share of Research Budget," RollCall.com, August 3, 2017.

28. Barney Frank, speaking to the Center for American Progress Action Fund, June 23, 2009.

29. See the discussion of Keynesianism in the preceding chapter.

30. Paul Krugman, "Decade at Bernie's," *NYT,* February 15, 2009.

31. Harry S. Truman, speech to Congress, March 12, 1947.

32. IHS Economics, "Aerospace and Defense Economic Impact Analysis," April 2016.

33. James Fallows, "The Tragedy of the American Military," *Atlantic,* January/February 2015.

34. "The Pentagon's Excess Space," *NYT* Editorial Board, February 7, 2015.

35. See Chalmers Johnson's trilogy: *Blowback* (2004); *The Sorrows of Empire* (2005); and *Nemesis* (2007).

36. "Study Shows that Domestic, Not Military Spending, Fuels Job Growth," *Brown University News*, May 25, 2017.

37. Joseph Stiglitz and Linda Bilmes, *The Three Trillion Dollar War: The True Cost of the Iraq Conflict*, 2008, citing the National Priorities Project.

38. Stuart W. Leslie, *The Cold War and American Science: The Military-Academic-Industrial Complex at MIT and Stanford*, 1993.

39. Seymour Melman, *Pentagon Capitalism: The Political Economy of War*, 1970.

40. Tom Engelhardt, *The American Way of War: How Bush's Wars Became Obama's*, 2010.

Chapter 12: A-Bombs and H-Bombs

1. The most comprehensive source of information on the US nuclear arsenal until near the end of the twentieth century is Chuck Hansen, *The Swords of Armageddon* (1995), an eight-volume, 2,500-page history of the US development of nuclear weapons.

2. Brookings Institution, "Estimated Minimum Incurred Costs of U.S. Nuclear Weapons Programs, 1940–1996."

3. Federation of American Scientists, "Status of World Nuclear Forces," FAS.org, May 2019.

4. Kyle Mizokami, "Asia's 5 Most Lethal Wars of All Time," *National Interest*, August 1, 2015.

5. Helen Caldicott, *The New Nuclear Danger*, 2004.

6. 111th Congress (2009–2010), "Summary of H.R. 2647: National Defense Authorization Act for Fiscal Year 2010," Section 1251.

7. See "Congress Increases Funding for Nuclear R&D in 2019," Nuclear Energy Institute, September 13, 2018.

8. "U.S. Nuclear Modernization Programs," Arms Control Association, March 2018.

9. The White House, Office of the Press Secretary, "Remarks by President Barack Obama in Prague as Delivered," April 5, 2009.

10. James N. Mattis, "Secretary's Preface," *Nuclear Posture Review*, US Department of Defense, February 2018.

11. See: Neil MacFarquhar and David E. Sanger, "Putin's 'Invincible' Missile Is Aimed at U.S. Vulnerabilities," *NYT*, March 1, 2018.

12. "US Adds 'Low Yield' Nuclear Weapon to Its Submarine Arsenal," *Associated Press*, February 4, 2020.

13. James Carroll, "How Many Minutes to Midnight?" TomDispatch, February 12, 2019.

14. Carroll, "How Many Minutes to Midnight?"

15. A "puny" ten-kiloton nuclear blast in Manhattan would cause an estimated 550,000 casualties. See "This Is What a Nuclear Bomb Looks Like," *New York Magazine*, June 11, 2018.

16. Michael Krepon, "The Folly of Tactical Nuclear Weapons," Defense One, October 2, 2017.

17. Herman Kahn, *On Escalation: Metaphors and Scenarios*, 1965.
18. Alex Wellerstein, citing Stephen I. Schwartz, *Atomic Audit: The Costs and Consequences of U.S. Nuclear Weapons Since 1940*, 1998.

Chapter 13: Non-nuclear Technologies of Death

1. Micheal Clodfelter, *Vietnam in Military Statistics: A History of the Indochina Wars, 1792–1991*, 1995. See also: THOR (Theater History of Operations), Air Force Research Institute, a database of every bomb the US military has dropped since World War I.
2. Tom Engelhardt, "For the 15 Years Since 9/11, the U.S. Has Waged an Endless Campaign of Violence in the Middle East," *Nation*, September 8, 2016.
3. "U.S.: Hundreds of Civilian Deaths in Iraq Were Preventable," Human Rights Watch, December 12, 2003.
4. "Combined Forces Air Component Commander 2007–2012 Airpower Statistics," Bureau of Investigative Journalism, October 31, 2012.
5. "Operation Inherent Resolve: Strike Update," US Department of Defense, August 9, 2017.
6. Harriet Agerholm, "Map Shows Where President Barack Obama Dropped His 20,000 Bombs," *Independent*, January 19, 2017.
7. Jennifer Wilson and Micah Zenko, "Donald Trump Is Dropping Bombs at Unprecedented Levels," *Foreign Policy*, August 9, 2017.
8. See Louis F. Fieser, *The Scientific Method: A Personal Account of Unusual Projects in War and in Peace*, 1964.
9. Quoted in Robert M. Neer, *Napalm: An American Biography*, 2013.
10. Curtis LeMay, *Mission with LeMay: My Story*, 1965.
11. Alan Rohn, "Napalm in Vietnam War," thevietnamwar.info, April 3, 2017.
12. Eleanor Jane Sterling, Martha Maud Hurley, and Le Duc Minn, *Vietnam: A Natural History*, 2006. This study by three wildlife specialists at the American Museum of Natural History describes in detail the lasting ecological effects of chemical warfare in Vietnam.
13. Quoted by James W. Crawley, "Officials Confirm Dropping Firebombs on Iraqi Troops," *San Diego Union Tribune*, August 5, 2003.
14. THOR (see note 1 above). For the scale of the bombing of Cambodia, see Ben Kiernan and Taylor Owen, "Iraq, Another Vietnam? Consider Cambodia," in Mark Pavlick and Caroline Luft, eds., *The United States, Southeast Asia, and Historical Memory*, 2019.
15. Fatima Bhojani, "Watch the U.S. Drop 2.5 Million Tons of Bombs on Laos," *Mother Jones*, March 26, 2014. See also: "About Laos," *Legacies of War*, 2018.
16. Joshua Kurlantzick, *A Great Place to Have a War: America in Laos and the Birth of a Military CIA*, 2017.
17. Quoted by Eni Faleomavaega, chairman of the US House of Representatives Subcommittee on Asia, the Pacific and the Global Environment, April 22, 2010. See "Chairman Eni Faleomavaega Statement," *Legacies of War*, 2018.
18. Lao National Regulatory Authority for UXO, cited in "Obama to Address Lethal Legacy of Secret War in Laos," *NYT*, September 5, 2016.

19. Santi Suthinithet, "Land of a Million Bombs," *Hyphen Magazine*, 2010.
20. Laos's UXO problem is the subject of an excellent book and documentary film, both entitled *Eternal Harvest: The Legacy of American Bombs in Laos*. See: eternalharvestthebook.com and eternalharvestfilm.com.
21. Rebecca Wright, "'My Friends Were Afraid of Me': What 80 Million Unexploded US Bombs Did to Laos," *CNN*, September 6, 2016.
22. "White House Fact Sheet: U.S.–Laos Relations," USAID.gov, March 21, 2017.
23. "Joint Declaration between the United States of America and the Lao People's Democratic Republic," White House, Obamawhitehouse.archives.gov, September 6, 2016.
24. Barack Obama, "Press Conference of President Obama after ASEAN Summit," White House, Obamawhitehouse.archives.gov, September 8, 2016.
25. See "U.S. Using Cluster Munitions in Iraq," Human Rights Watch, April 1, 2003; and "Cluster Bombs in Afghanistan," Human Rights Watch, October 2001.
26. Richard Kidd (director of the Office of Weapons Removal and Abatement), "Is There a Strategy for Responsible U.S. Engagement on Cluster Munitions?" US Department of State Archive, April 28, 2008.
27. Mary Wareham, "US Embraces Cluster Munitions," Human Rights Watch, December 1, 2017.
28. US Army, PE 0604802A: Weapons and Munitions Engineering Development Program, February 2018.

Chapter 14: Bombers, Missiles, and Antimissiles

1. Brookings Institution, *Atomic Audit: The Costs and Consequences of U.S. Nuclear Weapons Since 1940*, August 1998.
2. "B-2 Bomber: Cost and Operational Issues," US Government Accounting Office, August 1997. "Total program cost" includes research, development, engineering, and testing expenses.
3. Amanda Macias, "The Colossal Price to Fly a Pair of B-2 Bombers to Hit Two ISIS Camps in Libya," *Business Insider*, January 19, 2017.
4. Sebastien Roblin, "The Crazy Story of How the Stealth F-35 Fighter Was Born," *National Interest*, February 24, 2019. Emphasis added.
5. Andrea Drusch, "Fighter Plane Cost Overruns Detailed," quoting Marine Lt. Gen. Robert Schmidle, *Politico*, February 16, 2014.
6. Paul Barrett, "Is the F-35 a Trillion-Dollar Mistake?" *Bloomberg*, April 4, 2017.
7. John McCain, "Remarks by Senator John McCain on the 'Military-Industrial-Congressional' Complex," December 15, 2011.
8. Adam Silverman, "Pentagon F-35 Review Unlikely to Affect Vermont," *Burlington Free Press*, February 5, 2017.
9. "Lockheed Martin Meets 2018 F-35 Production Target with 91 Aircraft Deliveries," f35.com [a Lockheed-Martin website], December 20, 2018.
10. John McCain, "Remarks by Senator John McCain at the Marine Fighter Attack Squadron 121 Re-designation Ceremony," November 20, 2012.
11. Quoted in Barrett, "Is the F-35 a Trillion-Dollar Mistake?"
12. Barrett, "Is the F-35 a Trillion-Dollar Mistake?"

13. See the section in chapter 11 entitled "Weaponized Keynesianism."

14. See chapter 20, "Operation Paperclip: The Nazification of American Science."

15. "New START Treaty Aggregate Numbers of Offensive Arms," US State Department, September 1, 2017.

16. General Hyten, after being chosen in July 2019 by the Trump administration to be vice chairman of the Joint Chiefs of Staff, gained notoriety due to accusations by an army colonel that he had sexually assaulted her in 2017.

17. Jon Harper, "STRATCOM Chief Bashes Acquisition Trends for Nuclear Systems," *Breaking Defense,* June 20, 2017.

18. See the discussion of the RAND Corporation in chapter 21.

19. See the discussion of deterrent strategy in chapter 21.

20. Future plans for the Trident missile were discussed in chapter 12 in the section headed "'Little' Nukes—No Big Deal?"

21. Paul Iddon, "The Syria Strike Proves America's Military Is Addicted to Tomahawk Missiles," *National Interest,* April 10, 2017.

22. "Tomahawk Cruise Missile," Raytheon Company, 2018.

23. Rebecca Slayton, "The Fallacy of Proven and Adaptable Defenses," Federation of American Scientists, August 19, 2014.

24. Jon Krakauer, *Where Men Win Glory: The Odyssey of Pat Tillman,* 2009.

25. Spencer Ackerman, "41 Men Targeted but 1,147 People Killed: US Drone Strikes—The Facts on the Ground," *Guardian,* November 24, 2014.

26. Quoted in Sherry Michaels, *The Rise of American Air Power: The Creation of Armageddon,* 1989.

27. GAO, "Operation Desert Storm: Evaluation of the Air Campaign," June 12, 1997.

28. Gar Smith, "Bombs Awry! The Imprecision of 'Precision' Bombing," Environmentalists Against War, May 25, 2003.

29. R. Jeffrey Smith, "Hypersonic Missiles Are Unstoppable. And They Are Creating a New Global Arms Race." *NYT Magazine,* June 19, 2019.

30. Steven Simon, "Hypersonic Missiles Are a Game-Changer," *NYT,* January 2, 2020.

31. Smith, "Hypersonic Missiles Are Unstoppable."

32. Associated Press, "New Russian Weapon Can Travel 27 Times the Speed of Sound," *NYT,* December 27, 2019.

33. Smith, "Hypersonic Missiles Are Unstoppable."

34. Thomas W. Ray, "A History of the DEW Line, 1946–1965," ADC [Air Defense Command] Historical Study No. 31, June 1965.

35. DARPA (Defense Advanced Research Projects Agency) will be discussed in chapters 15, 16, and 17.

36. Max Fisher, *NYT,* January 14, 2018.

37. For a less superficial analysis, see John Tirman, "How We Ended the Cold War," *Nation,* October 14, 1999.

38. See chapter 9.

39. Quoted by Patrick Tyler in "How Edward Teller Learned to Love the Nuclear-Pumped X-ray Laser," *Washington Post,* April 3, 1983.

40. Barbara Carton, "Area Scientists Join Growing Protest, Pledge to Shun SDI Funds," *Washington Post,* October 15, 1986.

41. Peter Goodchild, *Edward Teller: The Real Dr. Strangelove*, 2004. For other "real" Dr. Strangeloves, see chapter 21.
42. Carton, "Area Scientists Join Growing Protest."
43. "The Dangerous Illusion of Missile Defense," *NYT* Editorial Board, February 11, 2018.
44. David E. Sanger and William Broad, "Trump Vows to Reinvent Missile Defenses, but Offers Incremental Plans," *NYT*, January 17, 2019.

Chapter 15: Video-Game War

1. The pilot, Matt Martin, was quoted in Chris Cole, "'Smart Weapons Systems': Are We Being Misguided about 'Precision Strikes'?" *Global Research*, December 5, 2015.
2. Center for the Study of the Drone at Bard College, *Drones in the Defense Budget*, October 2017.
3. "America's Forever Wars," *NYT* Editorial Board, October 22, 2017.
4. Eric Schmitt, "A Shadowy War's Newest Front: A Drone Base Rising from Saharan Dust," *NYT*, April 22, 2018.
5. John Sifton, "A Brief History of Drones," *Nation*, February 7, 2012.
6. For an analysis of this event, see Jeremy Scahill, "With Suleimani Assassination, Trump Is Doing the Bidding of Washington's Most Vile Cabal," *The Intercept*, January 3, 2020.
7. Jo Becker and Scott Shane, "Secret 'Kill List' Proves a Test of Obama's Principles and Will," *NYT*, May 29, 2012.
8. Rebecca Gordon, "Forget 'America First'—Donald Trump's Policy Is Drones First," *Nation*, May 25, 2018.
9. Spencer Ackerman, "Trump Ramped Up Drone Strikes in America's Shadow Wars," *Daily Beast*, November 26, 2018.
10. "The Secret Death Toll of America's Drones," *NYT* Editorial Board, March 30, 2019.
11. Kathryn Watson, "Trump Nixes Public Report on Civilians Killed by Drone Strikes," *CBS News*, March 6, 2019.
12. The video can be viewed on YouTube. Search for "Bug-Sized Lethal Drones Being Developed by U.S. Air Force."
13. Annie Jacobsen, *The Pentagon's Brain: An Uncensored History of DARPA, America's Top Secret Military Research Agency*, 2015.

Chapter 16: Lethal Autonomy

1. For a useful introduction to this subject, see Paul Scharre, *Army of None: Autonomous Weapons and the Future of War*, 2018.
2. Peter Finn, "A Future for Drones: Automated Killing," *Washington Post*, September 19, 2011.
3. Defense Science Board, Department of Defense, "The Role of Autonomy in DoD Systems," July 2012.
4. Matthew Rosenberg and John Markoff, "The Pentagon's 'Terminator Conun-

drum': Robots That Could Kill on Their Own," *NYT,* October 25, 2016.

5. US Department of Defense, *Unmanned Systems Integrated Roadmap FY2013–2038,* October 2014.

6. Kashmir Hill and Aaron Krolik, "How Photos of Your Kids Are Powering Surveillance Technology," *NYT,* October 11, 2019.

7. *Wall Street Journal,* April 3, 2018.

8. Charlie Warzel, "A Major Police Body Cam Company Just Banned Facial Recognition," *NYT,* June 27, 2019.

9. SyNAPSE: Systems of Neuromorphic Adaptive Plastic Scalable Electronics.

10. Annie Jacobsen, *The Pentagon's Brain: An Uncensored History of DARPA, America's Top Secret Military Research Agency,* 2015.

11. Rosenberg and Markoff, "Pentagon's 'Terminator Conundrum.'"

12. DARPA, "Personal Assistant that Learns," darpa.mil, undated.

13. Alex Davies, "Inside the Races that Jump-Started the Self-Driving Car," *Wired,* November 10, 2017.

14. General Services Administration, "Short-Range Independent Microrobotic Platforms (SHRIMP)," fbo.gov, undated.

15. *60 Minutes,* CBS News, April 12, 2009.

16. Jacobsen, *Pentagon's Brain.*

17. "How Much Does a Prosthetic Arm Cost?" Health.CostHelper.com, 2018. Veterans with service-related amputations, however, are promised "cutting-edge prosthetic technology" and "new prostheses as new technology becomes available" by the Veterans Health Administration.

18. Benjamin Lambeth, "Technology Trends in Air Warfare," RAND Corporation, 1996. See the discussion of "flybots" in the previous chapter.

19. Bruce Upbin, "First Look at a DARPA-Funded Exoskeleton for Super Soldiers," *Forbes,* October 29, 2014.

20. Liam Stoker, "Creating Supermen: Battlefield Performance Enhancing Drugs," *Army Technology,* April 14, 2013.

21. Stoker, "Creating Supermen."

22. The President's Council on Bioethics, October 2003.

23. Stoker, "Creating Supermen."

24. DARPA, "Biological Robustness in Complex Settings (BRICS)," ITgrants.info, September 2018.

25. Colin Clark, "DepSecDef on Boosted Humans & Robot Weapons," *Breaking Defense,* March 30, 2016.

26. DARPA, "DARPA and the Brain Initiative," darpa.mil, undated.

27. *Forbes,* July 10, 2017. For a cogent survey of DARPA's efforts to weaponize human brains, see Michael Joseph Gross, "The Pentagon's Push to Program Soldiers' Brains," *Atlantic,* November 2018.

28. Eliza Strickland, "DARPA Wants Brain Implants that Record from 1 Million Neurons," *IEEE Spectrum,* July 10, 2017.

29. Sydney J. Freedberg, "Pentagon Studies Weapons that Can Read User's Mind," *Breaking Defense,* July 14, 2017. Emphasis in original.

30. D.D. Schmorrow and A.A. Kruse, "DARPA's Augmented Cognition Program,"

IEEE Digital Library, September 19, 2002.

31. Quoted in Colin Clark, "The Terminator Conundrum," *Breaking Defense*, January 21, 2016.

32. Human Rights Watch and International Human Rights Clinic, *Losing Humanity: The Case against Killer Robots*, November 19, 2012.

33. "Autonomous Weapons: An Open Letter from AI and Robotics Researchers," FutureOfLife.org, July 2015.

34. See chapter 18.

Chapter 17: Is Cyberwarfare Really a Thing?

1. Sydney J. Freedberg, "'Cyberwar' Is Over Hyped," *Breaking Defense*, September 10, 2013.

2. White House cybersecurity aide Jason Healey, quoted in Freedberg, "'Cyberwar' Is Over Hyped."

3. Bolton's tenure as national security advisor ended with his dismissal on September 10, 2019.

4. David E. Sanger, "Trump Loosens Secretive Restraints on Ordering Cyberattacks," *NYT*, September 20, 2018. For a thorough analysis of US vulnerability to cyberattacks, see: David E. Sanger, *The Perfect Weapon*, 2018.

5. *Weapons System Cybersecurity*, US Government Accountability Office, October 2018.

6. The Columbia-class submarines are expected to replace the Ohio-class nuclear-armed submarines described in chapter 14.

7. The Ground Based Nuclear Deterrent is expected to replace the current fleet of Minuteman III ICBMs, as described in chapter 14.

8. All of the bullet-point items below are verbatim quotations from the GAO report.

9. *Weapons System Cybersecurity*, US GAO.

10. Page O. Stoutland and Samantha Pitts-Kiefer, *Nuclear Weapons in the New Cyber Age*, Nuclear Threat Initiative, September 2018.

11. "Significant Cyber Incidents," Center for Strategic and International Studies, October 2018.

12. Eben Moglen, "Privacy Under Attack: The NSA Files Revealed New Threats to Democracy," *Guardian*, May 27, 2014.

13. Glenn Greenwald, "The Crux of the NSA Story in One Phrase: 'Collect It All,'" *Guardian*, July 15, 2013. Emphasis in original. See also: Edward Snowden, *Permanent Record*, 2019.

14. Quoted in Gregor Peter Schmitz, "Ex-Präsident Carter Verdammt US-Schnüffelei," *Der Spiegel*, July 17, 2013.

15. Eric Lichtblau, "In Secret, Court Vastly Broadens Powers of N.S.A.," *NYT*, July 6, 2013.

16. Peter Baker and David E. Sanger, "Obama Calls Surveillance Programs Legal and Limited," *NYT*, June 7, 2013.

17. See: Jennifer Stisa Granick and Christopher Jon Sprigman, "The Criminal N.S.A.," *NYT*, June 27, 2013.

18. *Statistical Transparency Report, Regarding Use of National Security Authorities,*

Fiscal Year 2017, Office of the Director of National Intelligence, April 2018.

19. Shoshana Zuboff, *The Age of Surveillance Capitalism*, 2019, is a thorough exposition of the science of data mining and its devastating social consequences.

20. Explaining the qubit in natural language may not be possible, but for valuable insights into the phenomenon, see Scott Aronson, "Why Google's Quantum Computing Milestone Matters," *NYT*, October 30, 2019. Dr. Aronson is a leading quantum computing researcher.

21. For a brief popular introduction to quantum computing, see Dennis Overbye, "Quantum Computing Is Coming, Bit by Qubit," *NYT*, October 23, 2019.

22. Glenn S. Gerstell, "I Work for N.S.A. We Cannot Afford to Lose the Digital Revolution," *NYT*, September 10, 2019.

23. Aaron Stanley, "Is the U.S. Getting Its Act Together on Quantum Computing?," *Forbes*, June 26, 2018.

24. Gerstell, "I Work for N.S.A."

25. Cade Metz, "Google Claims a Quantum Breakthrough that Could Change Computing," *NYT*, October 23, 2019.

26. National Academies of Sciences, Engineering, and Medicine, *Quantum Computing: Progress and Prospects*, 2019.

27. Anna Mitchell and Larry Diamond, "China's Surveillance State Should Scare Everyone," *Atlantic*, February 2, 2018.

Chapter 18: American Exceptionalism and the Ultimate Perversion of the Behavioral Sciences

1. In a speech against the war in Vietnam at Riverside Church in New York City, April 4, 1967.

2. David Brooks, "A Return to National Greatness," *NYT*, February 3, 2017.

3. Maureen Dowd, "Trump's Pile of Rubble," *NYT*, August 10, 2019. Trump's extreme anti-immigrant demagogy and policies have deflected attention from his predecessor's woeful treatment of immigrants. Immigrants' rights groups called Barack Obama the "Deporter in Chief" for authorizing the deportation of more than three million people.

4. Noa Yachot, "Trump Embraces the Original Sin of Guantánamo," aclu.org, January 31, 2018.

5. Yachot, "Trump Embraces the Original Sin."

6. White House news conference, August 1, 2014.

7. The arduous struggle of the SSCI report's authors to overcome fierce CIA resistance to its being made public is depicted in a 2019 film, *The Report*.

8. US Senate Select Committee on Intelligence, *Report of the Central Intelligence Agency's Detention and Interrogation Program* [redacted summary], December 9, 2014.

9. Scott A. Allen, "Nuremberg Betrayed: Human Experimentation and the CIA Torture Program," Physicians for Human Rights, June 5, 2017.

10. SSCI Torture Report, p. 113.

11. CIA, "OMS Guidelines on Medical and Psychological Support to Detainee Rendition, Interrogation, and Detention," December 2004 (approved for release

June 10, 2016).

12. M. Gregg Bloche, "When Doctors First Do Harm," *NYT*, November 22, 2016.

13. Bloche, "When Doctors First Do Harm."

14. Alfred W. McCoy, "Torture at Abu Ghraib Followed CIA's Manual," *Boston Globe*, May 14, 2004.

15. McCoy, "Torture at Abu Ghraib Followed CIA's Manual."

16. McCoy, "Torture at Abu Ghraib Followed CIA's Manual."

17. *KUBARK Counterintelligence Interrogation*, July 1963. Excerpts are available at National Security Archive, George Washington University. "Kubark" was a code name for the CIA during the Vietnam War.

18. Alfred W. McCoy, *The CIA's Secret Research on Torture*, 2014.

19. For the Nazi antecedents of Dr. Beecher's work, see the section headed "The Nazi Nexus in the Behavioral Sciences" in chapter 20.

20. Quoted by Neil A. Lewis, "Red Cross Finds Detainee Abuse in Guantánamo," *NYT*, November 30, 2004.

21. Jane Mayer, "The Experiment," *New Yorker*, July 11, 2005.

22. Jane Mayer, *The Dark Side*, 2008.

23. Tom Blanton, "Gina Haspel CIA Torture Cables Declassified," National Security Archive, August 10, 2018. Haspel compounded her criminality by destroying ninety-two videotapes documenting the waterboarding and other tortures.

24. Quoted by Sheri Fink, "2 Psychologists in C.I.A. Interrogations Can Face Trial, Judge Rules," *NYT*, July 28, 2017.

25. The lawsuit was settled in August 2017. The terms of the settlement were confidential, but the ACLU described the outcome as "a historic victory for our clients and the rule of law." See: "CIA Torture Psychologists Settle Lawsuit," aclu.org, August 19, 2017.

26. Quoted in McCoy, *CIA's Secret Research on Torture*.

27. The abbreviation APA is used herein to designate the American Psychological Association, which is not to be confused with the American Psychiatric Association.

28. APA, "2008 APA Petition Resolution Ballot," apa.org.

29. Stephen Soldz et al., *All the President's Psychologists*, April 2015.

30. David H. Hoffman et al., *Independent Review Relating to APA Ethics Guidelines, National Security Interrogations, and Torture*, July 2015.

31. American Psychological Association press release, July 10, 2015.

32. Robert Jay Lifton, letter to the *NYT*, December 15, 2014. Lifton is the author of *The Nazi Doctors: Medical Killing and the Psychology of Genocide*, 1988.

Chapter 19: The Explosive Birth of Big Science

1. IG Farben was an industrial syndicate formed in 1925 by the merger of six large German chemical, pharmaceutical, and dyestuff firms. After World War II, the Allies dissolved it into its original constituents.

2. US Department of Energy, "Manhattan Project Background Information and Preservation Work," undated. $2.2 billion in 1945 dollars is equivalent to about $30 billion in 2017 dollars.

3. Richard Rowberg, "Federal R&D Funding: A Concise History," Congressional Research Service, August 14, 1998. (The 1947 amount has been converted from 1998 to 2016 dollars.)

4. Congressional Budget Office, *The Budget and Economic Outlook*, Appendix H: "Historical Budget Data," February 2014.

5. Irvin Stewart, *Organizing Scientific Research for War: The Administrative History of the Office of Scientific Research and Development*, 1948.

6. Michael Meyer, "The Rise and Fall of Vannevar Bush," *Distillations*, Science History Institute, July 21, 2018.

7. See: US Department of Energy, "A Tentative Decision To Build the Bomb," The Manhattan Project: An Interactive History, OSTI.gov, undated.

8. See the section headed "A Crucial Political Struggle that Shaped DARPA" below.

9. I have borrowed this phrase from Richard Creath, "The Unity of Science: Carnap, Neurath, and Beyond," in Galison and Stump, eds., *The Disunity of Science*, 1990.

10. Daniel S. Greenberg, *Science, Money, and Politics: Political Triumph and Ethical Erosion*, 2001.

11. For a cogent Vietnam-era analysis of the JASON group's activities, see "Hasten, Jason—Guard the Nation," *Science for the People* magazine, September 1972. For a more recent analysis, see Ann Finkbeiner, *The Jasons: The Secret History of Science's Postwar Elite*, 2007.

12. Michael Goldblatt, DARPA's director of Defense Sciences from 1999 to 2003, quoted in Annie Jacobsen, *The Pentagon's Brain: An Uncensored History of DAR-PA, America's Top Secret Military Research Agency*, 2015.

13. Ann Finkbeiner, "Jason—A Secretive Group of Cold War Science Advisers—Is Fighting to Survive in the 21st Century," *Science*, June 27, 2019.

14. See the section headed "Edward Teller and the Death Ray" in chapter 14.

15. Memorandum, General Advisory Committee, October 25, 1949. Quoted in Jacobsen, *Pentagon's Brain*.

16. As an introduction to the remarkable scientific accomplishments of Henrietta Swan Leavitt, see George Johnson, *Miss Leavitt's Stars: The Untold Story of the Woman Who Discovered How to Measure the Universe*, 2005.

17. Eli Kintisch, "DARPA to Explore Geoengineering," *Science*, March 14, 2009. For an incisive critique of the theory and practice of geoengineering, see *Science for the People*, Special Issue: "Geoengineering," Summer 2018.

18. Geochemist Ken Caldeira of the Carnegie Institution for Science, quoted in Kintisch, "DARPA to Explore Geoengineering."

Chapter 20: Operation Paperclip:
The Nazification of American Science

1. Annie Jacobsen, *Operation Paperclip: The Secret Intelligence Program that Brought Nazi Scientists to America*, 2014.

2. Linda Hunt, *Secret Agenda: The United States Government, Nazi Scientists, and Project Paperclip, 1945–1990*, 1991.

3. Jacobsen, *Operation Paperclip*.

4. Smithsonian National Air and Space Museum, "V-2 Missile."

5. "At least ten thousand" is technically true, but the US Army's Dora-Nordhausen war crimes trial in August 1947 charged the Nazis with working at least twenty thousand slave laborers to death there. See also the credible estimate of thirty thousand victims in Jean Michel, *Dora: The Nazi Concentration Camp Where Modern Space Technology Was Born and 30,000 Prisoners Died*, 1980.

6. Tom Lehrer, "Wernher von Braun," from the album *That Was the Week That Was*, 1965.

7. Linda Hunt, "U.S. Coverup of Nazi Scientists," *Bulletin of the Atomic Scientists*, 1985.

8. See note 5 above.

9. For a detailed account of the Dora facility and von Braun's role, see: André Sellier, *A History of the Dora Camp: The Story of the Nazi Slave Labor Camp that Secretly Manufactured V-2 Rockets*, 2003.

10. Quoted by Jacobsen, *Operation Paperclip*; emphasis added. Peenemünde, a small town on an island in the Baltic Sea, was the site of a Nazi research center.

11. April 12, 1946.

12. Jacobsen, *Operation Paperclip*. Much of the information on the Nazi scientists reported in this chapter is from this source.

13. Wolfgang Saxon, "Arthur Rudolph, 89, Developer of Rocket in First Apollo Flight," *NYT*, January 3, 1996. In the early 1980s, however, Rudolph's heinous war crimes at Dora had been exposed. Rather than face trial, Rudolph chose to forfeit his US citizenship and leave the country permanently.

14. William J. Broad, "Dr. Kurt Heinrich Debus Is Dead; Helped Develop Modern Rocketry," *NYT*, October 11, 1983.

15. Jacobsen, *Operation Paperclip*.

16. James E. Davis, ed., "NASA's Secret Relationships with U.S. Defense and Intelligence Agencies," National Security Archive, April 10, 2015.

17. Chris Miller, "The INF Treaty Is Dead, and Russia Is the Biggest Loser," *Foreign Policy*, August 2, 2019.

18. See: William D. Hartung, "Trump's Space Force Is Putting Us All in Danger," *Nation*, September 25, 2018.

19. Quoted in "President Trump's 'Space Force'," *ABC News*, August 7, 2018.

20. Neil deGrasse Tyson and Avis Lang, *Accessory to War: The Unspoken Alliance between Astrophysics and the Military*, 2018.

21. Tyson and Lang, *Accessory to War*.

22. Mark R. Campbell et al., "Hubertus Strughold, 'The Father of Space Medicine'," *Aviation, Space, and Environmental Medicine*, July 2007.

23. The private papers of Heinrich Himmler constitute the most important primary source of data on the inhumanity of Nazi research practices. See "Register of the Heinrich Himmler Papers," Hoover Institution, Online Archive of California.

24. Ralph Blumenthal, "Drive on Nazi Suspects a Year Later: No Legal Steps Have Been Taken," *NYT*, November 23, 1974.

25. Jacobsen, *Operation Paperclip*.

26. The facility was relocated to Brooks Air Force Base in San Antonio, Texas, in 1959.

27. See: Office of the Surgeon General, US Department of the Army, *Medical*

Aspects of Chemical and Biological Warfare, chapter 19: "The U.S. Biological Warfare and Biological Defense Programs."

28. Sarah Everts, "The Nazi Origins of Deadly Nerve Gases," *Chemical and Engineering News*, October 17, 2016. Information in this and the three following paragraphs is from this source.

29. Joint Intelligence Objectives Agency.

30. Jacobsen, *Operation Paperclip*. The internal quotation was from a letter to President Truman written by war crimes investigators Leo Alexander and Alexander Hardy.

31. "Ex-Nazi High Post with United States Air Force, says Medical Man Here," *Boston Globe*, December 9, 1951. Drew Pearson, "Air Force Hires Nazi Doctor Linked to Ghastly Experiments," *Fredericksburg (VA) Free Lance–Star*, February 14, 1952.

32. Sun Bean Kim et al., "Risk Factors for Mortality in Patients with *Serratia marcescens* Bacteremia," *Yonsei Medical Journal*, March 1, 2015.

33. The 1968 army report was titled "A Study of the Vulnerability of Subway Passengers in New York City to Covert Attack with Biological Agents."

34. National Academies of Sciences, Engineering, and Medicine, *Health Effects of Project Shad Biological Agent: Bacillus Globigii*, 2004.

35. The most thorough account of the entire secret bioweapon testing program is: Leonard Cole, *Clouds of Secrecy: The Army's Germ Warfare Tests over Populated Areas*, 1999.

36. *US Congressional Record*, April 23, 1997.

37. "Watchdog Group Votes to Punish Syria for Chemical Weapon Use," *Washington Post*, April 21, 2021. Although the evidence of Syrian chemical warfare cited by the OPCW seems credible to me, I cannot ignore the strong challenges to its veracity by investigative journalists whom I respect, including Seymour Hersh and Chris Hedges. I have no additional evidence to contribute on this subject.

38. Quoted in Alfred W. McCoy, *The CIA's Secret Research on Torture*, 2014.

39. McCoy, *CIA's Secret Research on Torture*.

40. See chapter 18.

41. McCoy, *CIA's Secret Research on Torture*.

42. Quoted in McCoy, *CIA's Secret Research on Torture*.

43. Jeffrey St. Clair and Alexander Cockburn, "The Abominable Dr. Gottlieb," *CounterPunch*, November 17, 2017.

44. Stephen Kinzer, *Poisoner in Chief: Sidney Gottlieb and the CIA Search for Mind Control*, 2019.

45. Stephen Kinzer, in an interview with Terry Gross, on National Public Radio, September 9, 2019.

46. Kinzer, *Poisoner in Chief*. For a more thorough account including extensive eye-witness evidence, see Hal Gold, *Japan's Infamous Unit 731: Firsthand Accounts of Japan's Wartime Human Experimentation Program*, 2019.

47. Quoted in Kinzer, *Poisoner in Chief*.

48. Nicholas Kristof, "Unmasking Horror—Japan Confronting Gruesome War Atrocity," *NYT*, March 17, 1995.

49. Among the earliest and best is Sheldon H. Harris, *Factories of Death: Japanese Biological Warfare, 1932–1945, and the American Cover-Up*, 1994.

Chapter 21: The RAND Corporation:
From "Fuck You, Buddy" to Doomsday

1. Chalmers Johnson, "A Litany of Horrors: America's University of Imperialism," TomDispatch, April 29, 2008.

2. Alex Abella, "The Rand Corporation: The Think Tank that Controls America," Mental Floss, June 30, 2009.

3. "Obituary: Bruno W. Augenstein," *Los Angeles Times*, July 17, 2005.

4. "The Neutron Bomb," Nuclear Age Peace Foundation, undated.

5. Hans M. Kristensen and Robert S. Norris, "Global Nuclear Weapons Inventories, 1945–2013," *Bulletin of the Atomic Scientists*, 2013.

6. Kristensen and Norris, "Global Nuclear Weapons Inventories."

7. Alex Abella, *Soldiers of Reason: The RAND Corporation and the Rise of the American Empire*, 2008.

8. John Foster Dulles, "The Evolution of Foreign Policy," Department of State, Press Release No. 81, January 12, 1954.

9. Louis Menand, "Fat Man: Herman Kahn and the Nuclear Age," *New Yorker*, June 27, 2005.

10. Menand, "Fat Man."

11. Stanley Kubrick, *Dr. Strangelove, or How I Learned to Stop Worrying and Love the Bomb*, 1964.

12. *A Beautiful Mind* won the 2002 Oscar for Best Picture, and Russell Crowe was nominated for Best Actor for his portrayal of John Nash.

13. See the sections in chapter 10 about James M. Buchanan.

14. The Prisoner's Dilemma: Imagine yourself and a colleague finding yourselves under arrest and charged with conspiracy against the state. You are held separately from your colleague and cannot communicate with her. Your captors want one or both of you to confess your crime and implicate the other. To that end, they offer one-year prison sentences if both of you insist on maintaining your innocence. But if one of you betrays the other, the betrayer is immediately released while the betrayee is sentenced to twenty years. The best outcome for both of you would be for each to maintain your innocence. But because you each have to decide alone, you have to consider that your colleague might confess and implicate you. Therefore, your most "rational" option—regardless of what she actually does—is for you to confess and implicate her.

15. M. Hausner, J. F. Nash, L. S. Shapley, and M. Shubik, "'So Long Sucker,' A Four-Person Game," in M. Shubik, ed., *Game Theory and Related Approaches to Social Behavior*, 1964.

16. "So Long Sucker (1964)," Board Game Database, April 17, 2016.

17. Marina von Neumann Whitman, an economic advisor to Richard Nixon, quoted by Annie Jacobsen, *The Pentagon's Brain: An Uncensored History of DARPA, America's Top Secret Military Research Agency*, 2015.

18. Benjamin Schwarz, "America's Think Tank," *Columbia Journalism Review*, May/June 2008.

19. A pejorative term the US military applied to the National Liberation Front combatants.

20. See: John C. Donnell, Guy J. Pauker, and Joseph J. Zasloff, "Viet Cong Motivation and Morale in 1964: A Preliminary Report," RAND Corporation, March 1965. The quoted summary is from Anthony Russo, "Looking Backward: RAND and Vietnam in Retrospect," *Ramparts*, November 1972.

21. Duang Van Mai Elliott, *RAND in Southeast Asia: A History of the Vietnam War Era*, 2010.

22. See: Nick Cullather, "Bomb Them Back to the Stone Age: An Etymology," History News Network, October 5, 2006.

23. Barbara Myers, "The Secret Origins of the CIA's Torture Program and the Forgotten Man Who Tried to Expose It," *Nation*, June 1, 2015.

24. RAND analyst Gus Shubert, quoted in Mai Elliott, *RAND in Southeast Asia*.

25. Leon Goure, A. J. Russo, D. H. Scott, "Some Findings of the Viet Cong Motivation and Morale Study," RAND Corporation, June–December 1965.

26. Myers, "Secret Origins of the CIA's Torture Program."

27. Mai Elliott, *RAND in Southeast Asia*.

28. Russo, "Looking Backward."

29. Alfred W. McCoy, "Confronting the CIA's Mind Maze," TomDispatch, June 7, 2009.

30. Daniel Ellsberg, *The Doomsday Machine: Confessions of a Nuclear War Planner*, 2017.

Chapter 22: Is Science-for-Human-Needs Possible?

1. With no apologies to Grover Norquist. See chapter 10 for the original quotation.

2. Loren Graham, *Science in Russia and the Soviet Union*, 1993.

3. NSF, *Science & Engineering Indicators*, 1998.

4. CIA *World Factbook*, 2018.

5. Richard Levins, "Progressive Cuba-Bashing," *Socialism and Democracy*, March 2005.

6. World Health Organization, "Physicians (per 1,000 People)—Country Ranking," undated.

7. See John M. Kirk and H. Michael Erisman, *Cuba's Medical Internationalism*, 2009.

8. Pol De Vos et al., "Cuba's International Cooperation in Health: An Overview," *International Journal of Health Services*, February 2007.

9. Associated Press, "Cuba Blasts US over End of Medical Missions in Some Nations," December 18, 2019.

10. Debra Evenson, *Medicc Review*, April 2018. The journal *Medicc Review* is a comprehensive source of information on Cuban medical science. See also: Andrés Cárdenas O'Farrill, "How Cuba Became a Biopharma Juggernaut," Institute for New Economic Thinking, March 5, 2018.

11. European Centre for Disease Prevention and Control, "Factsheet about Invasive Haemophilus Influenzae Disease," undated.

12. Bernie Sanders and Elizabeth Warren both advocated a "Medicare-for-all plan to nationalize the health insurance industry" (Jeff Stein, "Warren's 2020 Agenda: Break Up Monopolies, Give Workers Control over Corporations, Fight Drug Companies," *Washington Post*, December 31, 2018). See also Thomas Neuberger, "Bernie Sanders' Green New Deal Plan Will Nationalize Power Generation in the US," AlterNet, September 10, 2019.

13. Bob Marley, "War," a song adapted from a speech at the United Nations on October 4, 1963, by Ethiopian emperor Haile Selassie.

Epilogue: The COVID-19 Pandemic

1. D. Coggon, G. Rose, D. J. P. Barker, eds. *Epidemiology for the Uninitiated,* 2003.
2. "Spanish flu" is a misnomer. The evidence points to the virus originating in the United States. See "Scientists Learn History of Spanish Flu at Fort Riley," army.mil, May 19, 2017.
3. Estimates by the US Centers for Disease Control and Prevention (CDC). The world population has increased since 1918 from about 2 billion to about 7.5 billion. Comparable rates of infection and mortality today would mean that 2.5 billion people would catch the disease and 200 million would die from it.
4. Matt Apuzzo and David D. Kirkpatrick, "Covid-19 Changed How the World Does Science, Together," *NYT,* April 1, 2020.
5. Lee Fang, "Banks Pressure Health Care Firms to Raise Prices on Critical Drugs, Medical Supplies for Coronavirus," The Intercept, March 19, 2020.
6. Sharon Lerner, "Cronyism and Conflicts of Interest in Trump's Coronavirus Task Force," The Intercept, February 29, 2020.
7. Dr. Ashish Jha, in an interview on BBC World News America, March 16, 2020.
8. "Remarks by President Trump at Signing of the Coronavirus Preparedness and Response Supplemental Appropriations Act, 2020," whitehouse.gov, March 6, 2020.
9. "Remarks by President Trump, Vice President Pence, and Members of the White House Coronavirus Task Force in Press Briefing," whitehouse.gov, March 9, 2020.
10. Beth Cameron, quoted in "Trump Disbanded NSC Pandemic Unit that Experts had Praised," Associated Press, March 14, 2020.
11. Emily Baumgaertner and James Rainey, "Trump Administration Ended Pandemic Early-Warning Program to Detect Coronaviruses," *Los Angeles Times,* April 2, 2020.
12. Rosie McCall, "From 'Smoking Doesn't Kill' to Conversion Therapy—Mike Pence's Most Controversial Science Remarks," *Newsweek,* February 27, 2020.
13. Derrick Z. Jackson, "Coronavirus Pandemic: Science Sidelined in Trump Rose Garden Fiasco," Union of Concerned Scientists, March 16, 2020.
14. Lena H. Sun, "CDC to Cut by 80 Percent Efforts to Prevent Global Disease Outbreak," *Washington Post,* February 1, 2018.
15. Roni Caryn Rabin, Knvul Sheikh and Katie Thomas, "As Coronavirus Numbers Rise, C.D.C. Testing Comes Under Fire," *NYT,* March 10, 2020.
16. Peter Whoriskey and Neena Satija, "How U.S. Coronavirus Testing Stalled: Flawed Tests, Red Tape and Resistance to Using the Millions of Tests Produced by the WHO," *Washington Post,* March 16, 2020.
17. Lena H. Sun, "CDC Director Warns Second Wave of Coronavirus Is Likely To Be Even More Devastating," *Washington Post,* April 21, 2020.
18. Rob Wallace, *Big Farms Make Big Flu: Dispatches on Infectious Disease, Agribusiness, and the Nature of Science,* 2016.
19. For the complexities of RNA viruses' "proofreading" capabilities, see David

Cyranoski, "Profile of a Killer: The Complex Biology Powering the Coronavirus Pandemic," *Nature*, May 4, 2020.

20. Wallace, *Big Farms Make Big Flu*. Wallace credited microbiologist Kennedy Shortridge as a pioneering source of the virological insights he was imparting.

21. "Egg Industry Cruelty Cracked Wide Open," eggabuse.com, undated.

22. "Egg labels are often intentionally deceiving. . . . more than 20,000 birds can be severely confined in a windowless, warehouse-style shed and still have their eggs sold under 'free-range' or 'cage-free' labels." humanefacts.org, undated.

23. Tyson Foods is "the world's largest processor and marketer of chicken meat" (Wallace, *Big Farms Make Big Flu*). The CP Group is a private Thai conglomerate with investments in thirty countries and more than three hundred thousand employees.

24. D. Goodman and M. J. Watts, eds, *Globalising Food: Agrarian Questions and Global Restructuring*, 1997. Cited by Wallace, *Big Farms Make Big Flu*.

25. Wallace, *Big Farms Make Big Flu*. Emphasis added.

26. Rob Wallace, Alex Liebman, Luis Fernando Chaves, and Rodrick Wallace, "COVID-19 and Circuits of Capital," *Monthly Review*, March 27, 2020.

27. Wallace, Liebman, Chaves, and Wallace, "COVID-19 and Circuits of Capital." Deforestation is but one of the ecological complexities in the chain of pandemic causation these authors discuss.

28. Rob Wallace, "Notes on a Novel Coronavirus," *Monthly Review*, January 29, 2020.

29. Mike Davis, "In a Plague Year," *Jacobin*, March 14, 2020.

30. Martin Kaste, "U.S. Hospitals Prepare for a COVID-19 Wave," National Public Radio, March 6, 2020.

31. Davis, "In a Plague Year." Emphasis added.

32. ARC Advisory Group, arcweb.com, March 18, 2020.

33. Quartz, qz.com, March 22, 2020.

34. *The American Prospect*, March 18, 2020.

35. *Common Dreams*, March 23, 2020.

36. *Daily Sabah* [Turkey], March 17, 2020.

37. Farhad Manjoo, *NYT*, March 25, 2020.

Index

academic-industrial complex
  agribusiness and, 110, 113
  Big Tech and, 113
  Commerce Department and, 112
  consequences of, 109, 115–116
  ethics and, 108, 115
  extent of, 112
  fossil fuel industry and, 111, 114
  Koch, Charles and David and,
    121–122
  militarization of science and, 108
  pharmaceutical industry and,
    109–110, 113–114
  private funding and, 109–112, 115
  rise of, 108
  universities and
    Carnegie Mellon University and,
      199
    Florida Atlantic University and,
      121
    Harvard University and, 113
    Purdue University and, 110–111
    Rensselaer Polytechnical Institute
      and, 111–113, 116
    Seattle Pacific University and, 121
    Stanford University and, 121
    University of California and, 114
    University of California at
      Berkeley and, 113–114
    University of California at
      Berkeley and, 113–114
    University of Illinois and, 114

Afghanistan, xv–xvi, 161–162, 164,
  166–167, 176, 188–190, 193,
  217–218
Agent Orange, 87, 164
agribusiness
  academic-industrial complex and,
    12, 113
  China and, 23, 287
  COVID-19 and other pathogens
    and, 285–287, 292
  GMOs and. *See* Genetically
    Modified Organisms (GMOs)
  Green Revolution and, 20–24
  lobbying and, 32–33
agronomy, 10, 20–25
Albright, Madeleine, 142, 146
Amazon (corporation), 3, 197, 212
American Enterprise Institute (AEI),
  84, 119, 121, 146
American Exceptionalism, 7, 205,
  215–219
anticommunism, 43, 85, 129, 131,
  149–150, 166, 232. *See*
  also McCarthyism
antiscience, 4, 38–39, 41–44, 55,
  64–66, 76, 123–124, 283. *See*
  also climate change, denial of;
  vaccines, anti-vax movement
Apple, 3, 197
artificial intelligence, 3, 195–200,
  205. *See also* lethal autonomy
atomic bomb. *See* nuclear weapons

About Haymarket Books

Haymarket Books is a radical, independent, nonprofit book publisher based in Chicago.

Our mission is to publish books that contribute to struggles for social and economic justice. We strive to make our books a vibrant and organic part of social movements and the education and development of a critical, engaged, international left.

We take inspiration and courage from our namesakes, the Haymarket martyrs, who gave their lives fighting for a better world. Their 1886 struggle for the eight-hour day—which gave us May Day, the international workers' holiday—reminds workers around the world that ordinary people can organize and struggle for their own liberation. These struggles continue today across the globe—struggles against oppression, exploitation, poverty, and war.

Since our founding in 2001, Haymarket Books has published more than five hundred titles. Radically independent, we seek to drive a wedge into the risk-averse world of corporate book publishing. Our authors include Noam Chomsky, Arundhati Roy, Rebecca Solnit, Angela Y. Davis, Howard Zinn, Amy Goodman, Wallace Shawn, Mike Davis, Winona LaDuke, Ilan Pappé, Richard Wolff, Dave Zirin, Keeanga-Yamahtta Taylor, Nick Turse, Dahr Jamail, David Barsamian, Elizabeth Laird, Amira Hass, Mark Steel, Avi Lewis, Naomi Klein, and Neil Davidson. We are also the trade publishers of the acclaimed Historical Materialism Book Series and of Dispatch Books.

About the Author

PHOTO CREDIT: JEFF FASANO PHOTOGRAPHY

CLIFFORD D. CONNER taught history of science at the School of Professional Studies, CUNY Graduate Center. He is the author of *A People's History of Science* (Bold Type Books, 2005) and biographies of three revolutionaries: Jean-Paul Marat, Arthur O'Connor, and Colonel Edward Marcus Despard.